Springer Climate

Series Editor

John Dodson ⓘ, Chinese Academy of Sciences, Institute of Earth Environment, Xian, Shaanxi, China

Springer Climate is an interdisciplinary book series dedicated to climate research. This includes climatology, climate change impacts, climate change management, climate change policy, regional climate studies, climate monitoring and modeling, palaeoclimatology etc. The series publishes high quality research for scientists, researchers, students and policy makers. An author/editor questionnaire, instructions for authors and a book proposal form can be obtained from the Publishing Editor.

Now indexed in Scopus® !

More information about this series at http://www.springer.com/series/11741

Kodoth Prabhakaran Nair

Combating Global Warming

The Role of Crop Wild Relatives for Food
Security

 Springer

Kodoth Prabhakaran Nair
International Agricultural Scientist
Calicut, Kerala, India

ISSN 2352-0698 ISSN 2352-0701 (electronic)
Springer Climate
ISBN 978-3-030-23036-4 ISBN 978-3-030-23037-1 (eBook)
https://doi.org/10.1007/978-3-030-23037-1

This Springer imprint is published by the registered company Springer Nature Switzerland AG
The registered company address is: Gewerbestrasse 11, 6330 Cham, Switzerland

This book, written under very trying circumstances, is dedicated to my wife, Pankajam, a Nematologist trained in Europe, but, one who gave up her profession, and, instead, chose to be a home maker, more than four decades ago, when we had our son and daughter. She is my all, and, she sustains me in this difficult journey, that life is

India's great President late Dr. A. P. J. Abdul Kalam launching the book "**ISSUES IN NATIONAL AND INTERNATIONAL AGRICULTURE**", authored by Prof. Kodoth Prabhakaran Nair, in Raj Bhavan, Chennai, Tamil Nadu, India

Acknowledgements

It would be impossible to list all who have helped me while compiling this book. But, I would like to add, without fail, a few words of genuine appreciation for the inspiring guidance that Ms. Margaret Deignan, Senior Editor, Springer Nature, gave me, which was a source of great encouragement to me to embark upon this difficult book project which has much relevance in the global environmental context—the climate change.

Addendum

During the final stages of completing this manuscript, we lost Charlie, our beloved canine pet, due to a brief illness in the evening of February 24, 2019. The joy he brought us, during his life, is irreplaceable. May his soul rest in peace.

Contents

Chapter 1
Introduction

Global warming is a reality man has to live with. This is a very important issue
to recognize, because, of all the parameters that affect human existence, on planet
earth, it is the food security that is of paramount importance to life on earth and
which is most threatened by global warming. Future food security will be depen-
dent on a combination of the stresses, both biotic and abiotic, imposed by climate
change, variability of weather within the growing season, development of cultivars
more suited to different ambient conditions, and, the ability to develop effective
adaptation strategies which allow these cultivars to express their genetic potential
under the changing climate conditions. These may appear as challenges which may
be impossible to address because of the uncertainty in our ability to predict future
climate. However, these challenges also provide us the opportunities to enhance our
understanding of soil–plant–atmosphere interaction and how one could utilize this
knowledge to enable us achieve the ultimate goal of enhanced food security across
all areas of the globe.

Those plant species which are very closely related to field crops, including their
progenitors, which have the potential to contribute beneficial traits for crop improvement, such as, resistance to an array of biotic and abiotic stresses, and to enrich the
gene pool, leading ultimately to enhanced plant yield, thereby aiding humanity's
relentless search for production of more food to meet the ever growing needs of a
burgeoning population, are called "Crop Wild Relatives" (CWRs). In fact, CWRs
are known to have tremendous potential to sustain and enhance global food security,
thereby contributing enormously to humanity's well-being. Therefore their search,
characterization, and conservation in crop breeding programs assume great importance. Viewed against the recent global climate upheavals in global climate change,
the task becomes all the more important. Against the background of the disastrous
after effects, especially the alarming environmental hazards of the highly soil extractive farming, euphemistically known as the "green revolution", of the 1960s, the
task assumes much cruciality. Global warming is a real threat to humanity *vis-a-
vis* crop production. Anthropogenic release of green house gases is fast affecting
climate change leading to global warming. This is bound to increase soil organic
matter decomposition and aggravate soil water deficits in the years to come. Greater

© Springer Nature Switzerland AG 2019
K. P. Nair, *Combating Global Warming*, Springer Climate,
https://doi.org/10.1007/978-3-030-23037-1_1

frequency of high intensity rainfall events, run off and flooding in future cause excessive soil erosion losses unless offsetting conservation measures are taken. Mitigation strategy of diverting carbon di oxide from the atmosphere to soil organic carbon by adopting conservative agricultural practices of reduced or no tillage, crop residue retention and diverse cropping systems is now recognized which has the potential to offset CO_2 saturation in the atmosphere. But, these measures have only limited scope, as there are many inter regional and international constraints on embarking such a mode of action.

The International Panel on Climate Change (2014) had predicted dramatic changes in climate pattern in the current century and this will, inevitably, adversely affect crop production. If one juxtaposes the global population increase with changes in ambient temperature, a grim scenario emerges. Current projections suggest that global temperature will rise by 1.8–4 °C and the corresponding global population increase might touch 10 billion by 2020—an increase of 142.9% from the current level of ± 7 billion, which works to about an annual increase of 1.8%.

There are two kinds of "greenhouse effects", the natural which makes ambient temperature of some regions at high altitude hospitable for life forms, while, the more common one, extremely hazardous, occurring in the atmosphere due to emission of industrial gases, such as CO_2, and those from agricultural practices, like excessive and unbridled use of nitrogen fertilizers, such as urea, leading to huge accumulation of nitrous oxide (N_2O) in the atmosphere, which is lethal to both human habitation and plant life. Automobile emission is an important one contributing hugely to CO_2 concentration in the atmosphere, which has changed from the preindustrial level of 20 ppm (parts per million) to 40 ppm, which is a 2000% increase. This is, indeed, a phenomenal increase. But, the more worrying is the production of N_2O, commonly known as the "laughing gas" (Dinitrogen monoxide), a major scavenger of stratospheric ozone leading to global warming. Nitrous oxide is ranked third behind CO_2 and methane (CH_4) in its effect on global warming as a "Greenhouse Gas (GHG)". It is 310 times more effective in its heat trapping capacity compared to CO_2. Average life time of N_2O is 120 years. 30% of N_2O in the atmosphere is due to agricultural practices, primarily application of excessive nitrogenous fertilizers, like urea. A steep ramp-up of N_2O in the atmosphere coincided with the green revolution of the 1960s. When high food production is targeted through excessive use of N fertilizers, like urea, soil microbes convert N in the fertilizer urea into N_2O at a faster rate than normal. This has been a major factor leading to global warming with all its attendant adverse environmental fallout, like low precipitation, soil degradation, ground water depletion, etc. The Punjab State in India, known as the "Cradle" of the green revolution, where thousands of acres of land have been ruined beyond repair is a living testimony to this monumental environmental disaster.

The growing concern over the potentially devastating impacts of climate change on biodiversity and food security, when juxtaposed with burgeoning global population, implies that immediate measures be taken to conserve the CWRs and derive from them potentially usable genes to enhance crop yield. CWRs are a key tool to address the limits of genetic variation in domestic crops to adapt them to climate change. However, extension of their conservation and promotion of more systematic exploitation are hindered by a lack of understanding of their current and potential value, their diversity, and practically how they might be conserved.

Climate change and bird migration:

Climate change is turning Israel into a permanent winter ground for some 500 million migrating birds, which used to stop over briefly before flying over to warm plains of Africa, Israeli experts say. The birds now prefer to stay longer in cooler areas rather than cross into Africa, where encroaching deserts and frequent drought have made food a scarce commodity. "In the past decades, Israel has become more than just a short stopover for the birds because many more birds and a greater number of species can no longer cross the desert" reported ornithologist Shay Agmon, avian coordinator for the wetlands part of Agamon Hula in Northern Israel. "They will stay here for longer periods and eventually the whole pattern of migration will change". Cranes are one of the most abundant species to visit the Hula wetlands and Agmon reported that the number that prefer to stay in Israel until the end of March has risen from <1000 in 1950s to 45,000 currently. Although migrating birds are an attraction for tourists and ornithologists, their hunger for food from crop fields makes them a menace to farmers. Workers at the Hula Reserve which lies in the Syrian-African Rift valley have lured the birds from surrounding fields by feeding them at the wetlands and offering them a more comfortable existence. "It's harder for the birds to cross a much larger desert and they just can't do it. There is not enough fuel, there are not enough "gas stations" on the way, so, Israel has become their biggest "gas station", their biggest "restaurant", reported Agmon". The pattern of bird migration has much to learn from, for us, humankind, in its search for more food through hitherto unexplored avenues.

Greenland Ice Melt:

In the context of global warming, it is but relevant to discuss, if only in brief, the phenomenon of Greenland Ice Melt as a consequence of global warming, which will have a direct effect on human existence, *vis-à-vis* food production. Greenland ice sheet is approaching a melting threshold faster than scientists thought. In two decades, it would become a major contributor to sea level rise.

If all of Greenland's ice were to melt, it would raise sea levels by 23 ft, submerging some coastal cities. That would put everything south of West Palm Beach, Florida under water. Between 2003 and 2013, the rate of Greenland's ice loss has quadrupled, according to a new study, most of which is caused by global warming. Other recent research has revealed that oceans are warming 40% faster than experts thought and that Antarctic ice is melting six times faster than that was the case in the 1980s.

Greenland's ice is melting four times faster than it was 16 years ago. In 2012, Greenland lost more than 400 billion tons of ice—almost a fourfold than that lost in 2003. Except for one year lull between 2013 and 2014, those losses continue to accelerate.

A recent study used data from satellites and a GPS network on the ground to determine that Greenland's ice is melting faster than one thought. The authors of the research paper wrote that within the next 20 years, that melt "will become a major future contributor to sea-level rise." What is more, the study highlighted the risks in Greenland's southwestern region which is not typically known to be a source of ice loss. Most melt comes from other areas, where icebergs slough off glaciers and float out to sea. Greenland's southwest does have many such glaciers, but, it is responsible for more meltwater in the ocean than other parts of the island. "This is going to cause additional sea-level rise," said Michael Bevis, lead author of the research paper and a Professor at Ohio University, who told the National Geographic magazine "We are watching the ice-sheet hit a tipping point." This news comes in the wake of another ominous study published recently which found that Antarctica's ice melt is also speeding up. In the 1980s, Antarctica lost 40 billion tons of ice annually. In the last decade, that number jumped to an average of 252 billion tons per year. Recent research has also shown that oceans are heating up 40% faster than experts thought, and that 2018 was the warmest year on record for ocean temperatures. These findings need no further emphasis to point out what the implications of all of the above would be on global food production.

What happens if the entire Greenland ice sheet melts?

Roughly, 1.7 million sq.km (65,000 sq.miles) in size, the Greenland Ice Sheet covers an area almost three times the size of Texas. Together with Antarctica's ice sheet, it contains more than 99% of the world's fresh water, according to the National Snow and Ice Data Center. Currently, most of that water is frozen in masses of ice and snow that can be up to 10,000 ft thick. But, as we send more greenhouse gases into the atmosphere, the oceans absorb a lot of the excess heat they trap. And the warm air and water is lending frozen ice sheets to melt at unprecedented rates.

If the entire Greenland ice sheet were to melt—granted, this would take place over centuries—it would mean 23-feet sea level rise, on average. That is enough water to cover and submerge the entire southern tip of Florida.

If both Antarctica's and Greenland's ice sheets were to melt, that would lead sea level to rise over 200 ft and Florida would disappear. NASA has created an interactive tool which helps track sea level rise projections, based on how much the two ice sheets are melting. One thing the tool makes clear iscoastal cities will be heavily impacted. A map from National Geographic magazine shows how, in the event of a full ice melt, cities like Amsterdam, Stockholm, Buenos Aires, Dakar, and Cancun, to name just a few, would simply vanish from the face of the globe.

The above scenario is grim, and, must send a warning signal for the humankind to immediately initiate measures to lessen the rate of global warming. The million dollar question is, who will take the first step?

The prediction of a climate scientist, *par excellence* which has come true:

The name, Wallace Smith Broecker, may not be familiar to many, but, Dr. Broecker, the longtime Columbia University Professor was the one who coined the phrase "Global Warming" and made it a household term. He coined the phrase "Ocean Conveyor Belt", a global network of currents affecting everything from air temperature to rain patterns. "Wally was unique, brilliant and combative," said Princeton University Professor Michael Oppenheimer, mourning his death, after a long illness, in the second week of February 2019, at the ripe age of 87. "He was not fooled by the cooling of the 1970s. He saw clearly the unprecedented warming now playing out and made his views clear, even when few were willing to listen," said Professor Oppenheimer.

In the Ocean Conveyor Belt, cold, salty water in the North Atlantic sinks, working like a plunger to drive an ocean current from near North America to Europe. Warm surface waters borne by this current help keep Europe's climate mild. Otherwise, he said, Europe would be a deep freeze, with average winter temperatures dropping 20 °F or more and London feeling more like Spitsbergen, in Norway, which is 600 miles north of the Arctic Circle.

Broecker said his studies suggested that the Ocean Conveyor Belt is the "Achilles Heel of the Climate System and a fragile phenomenon which can change rapidly for reasons not clearly understood." It would take only a slight rise in temperature to keep water from sinking in the North Atlantic, he said, and that would bring the conveyor to a halt. Broecker said it is possible that warming caused by the build up of greenhouse gases could be enough to affect ocean currents dramatically. Broecker single-handedly popularized the notion that this could lead to a dramatic climate "tipping point" and, more generally, Broecker helped communicate to the public and policy makers the potential for abrupt climate changes and unwelcome "surprises" as a result of climate change, said Penn State Professor Michael Mann.

In 1984, Broecker told a US House Sub Committee that the build up of greenhouse gases warranted "a bold new national support aimed at understanding the operations of the realms of the atmosphere, oceans, ice and terrestrial biosphere".

"We live in a climate system that can jump abruptly from one state to another" told Broecker to the Associated Press in 1997. By dumping into the atmosphere huge amounts of greenhouse gases, such as carbon dioxide from the burning of fossil fuels, "we are conducting an experiment that could have devastating effects." "We are playing with an angry beast—a climate system that has been shown to be very sensitive," Broecker said.

Broecker received the National Medal of Science in 1996 and was a Member of National Academy of Science. He also served as the research coordinator for Biosphere 2, an experimental living environment turned research lab.

Evidence of man-made global warming hits "Gold Standard":

Scientists said that human activities were raising the heat at the earth's surface had reached a "five sigma" level, a statistical gauge meaning there is only a one-in-a-million chance that the signal would appear if there was no warming.

Evidence for man-made global warming has reached a "gold standard" level of certainty, adding pressure for cuts in greenhouse gases to limit rising temperatures, said scientists on February 25, 2019.

"Humanity cannot afford to ignore such clear signals" the US-led Team wrote in the journal Nature Climate change of satellite measurements of rising temperatures over the past 40 years. Such a "gold standard" as mentioned above was applied in 2012, for instance, to confirm the discovery of the Higgs boson subatomic particle, a basic building particle block of the Universe."

Benjamin Santer, lead author of the February 25, 2019 study at the Lawrence Livermore National Laboratory in California, said, he hoped the findings would win over skeptics and spur action. "The narrative out there that scientists don't know the cause of climate change is wrong," Santer told Reuters. His rejoinder was "We do".

Mainstream researchers say that the burning of fossil fuels is causing more floods, droughts, heat waves, and rising sea levels.

President Donald Trump has often cast doubt on global warming and plans to pull out of the 197-Nation Paris Climate Agreement, which seeks to end the fossil fuel era this century by shifting to cleaner energies like solar or wind. 62% Americans polled in 2018 believed that climate change has a human cause, up from 47% in 2013, according to the Yale Program on Climate Change Communication.

Satellite Data:

The February 25, 2019 findings, by researchers in the U.S, Canada, and Scotland, said the evidence for global warming reached the five sigma level by 2005 in two of three sets of satellite data, widely used by researchers, and in three sets by 2016. Separately in 2013, the U.N. Intergovernmental Panel on Climate Change (IPCC) concluded that it is extremely likely, or at least 95% probable, that human activities have been the main cause of climate change since the 1950s. Peter Scott of the British Met office, who was among the scientists drawing that conclusion and was not involved in the 25 February 2019 study, said he would favor rising probability one notch" to "virtually certain" or 99–100%. "The alternative explanation of natural factors dominating has got even less likely," he told Reuters.

The last four years have been the hottest since records began in the 19th century. The IPCC will next publish a formal assessment of the probabilities in 2021. "I would be reluctant to raise to 99–100%, but, there is no doubt there is more evidence of change in the global signals over a wider suite of ocean indices and atmospheric indices," said Professor Nathan Bindoff, a climate scientist in the University of Tasmania.

Climate change may cause Iceberg, twice the size of New York City, to break off Antarctic Ice shelf:

According to Skymetweather.com, climate change has been causing enough problems as far as melting icebergs and glaciers across Antarctica are concerned. In a new finding by NASA, an iceberg which is nearly twice the size of New York city is all ready to break away from an ice shelf of Antarctica. NASA says that a crack had appeared along the Brunt ice shelf in October 2016 which is now spreading over toward the East. The crack is also known as the Halloween crack, and is expected to interact with another fissure which was stable for about 3.5 decades, but is now moving north at lightning speed around 2.5 miles a year. When both of them will meet, which may happen in weeks, an iceberg of at least 660 sq.miles will break off. This process is called calving, occurs naturally with ice shelves; however, these changes are quite unfamiliar in this area. NASA has warned that it could cause destabilization of the Brunt ice shelf, leading to complete collapse as well. This would further accelerate ice formation in the upstream glaciers, which may cause an increased contribution to the sea level rise.

When the other side of Halloween Crack loses ice, instability will only increase. While this will be the largest from the Brunt Ice Shelf in ten decades, however, it is not the first time such an event to occur. NASA also says that this iceberg would not even make to the list of the 20 biggest icebergs in Antarctica. Moreover, in the year 2017, in July, icebergs of 2,200 sq.miles had calved from the Larsen C ice shelf. These icebergs were even bigger, twice the size of the State of Delaware.

Antarctic ice shelves will play a major role in global sea level rise in the future. US and UK scientists have claimed in 2018 that Antarctica ice melt is at a record-breaking rate, which may cause a major threat to coastal cities. Iceberg calving, although is natural, but due to climate change, ice shelves of Antarctica are thinning.

Ice sheet melting has paced up three times than usual in the last five years. This will only take a backseat if measures are taken up immediately to reduce greenhouse gas emissions and do something positive and proactive regarding global warming. But that, most unfortunately, seems not happening. By 2070, scientists say that melting ice in Antarctica should add more than 25 cm to total global sea level rise.

Ocean heat waves remake Pacific and Caribbean Habitats:

The short extreme heat wave may have a bigger impact than a slow warming. The colored corals have been bleached due to the heat waves and lost their photosynthetic symbiotes.

Climate change tends to deal in averages. We measure its progress using the global mean temperature and we use climate models to project what that value will be in the future. But those average changes do not always capture what future climate change will be like. While an average can be raised by increasing every day's temperature by a tiny amount, it is also possible to raise an average by throwing in an occasional extreme event. Heat waves and extreme storms have indicated that nature seems to be going for the latter option.

A new research paper that this kind of climate change is not just affecting the sorts of weather we typically experience, but, it is happening in the oceans as well. The study shows that, over the course of less than a century, the frequency of oceanic heat waves went up by more than 50%. The study looked into the effects these events are having on ecosystems, and, it showed that we are pushing species toward the poles without affecting all of them equally.

Heating the ocean's waves:

When the subject is the atmosphere, the common practice is to track the frequency and extent of heat waves and events to determine if they have been influenced by climate change. By contrast, there was no widely accepted definition of when warming waters constituted a heat wave until 2016. That is in part because of the differences in the driving process and scale. Localized ocean heat waves can be driven by a corresponding heat wave in the atmosphere, while El Nino events are driven by large-scale current patterns that influence most of the Pacific. But, the recent definition can encompass both of those. That is because it defines heat waves as exceeding a seasonal adjusted threshold (for example, hotter than 90% of the typical readings in spring) for at least five days. This not only provides a "yes" or "no" answer to "Is it a heat wave?", but, it also helps define the geographic extent of the extreme water temperatures, which can, in turn, help us understand the process driving it.

The authors took this definition and examined the historic record to track the frequency of marine heat waves (others have done a similar analysis and produced equivalent results). As one would expect on a warming planet, their results have gotten far more common. "As a global average, there were over 50% more (marine heat wave) days per year in the last part of the instrumental record (1987–2016) compared to the earlier part."

But that language obscures just how dramatic the change has been. Strong El Nino events over the last two decades have caused more heat waves than any in the historic record, and the recent one was roughly twice as strong as any prior to it. Many of the events are concentrated along the areas most affected by El Ninos, and this is a global problem, with very few areas of the oceans unaffected.

Swimming with the fishes:

This is not the first analysis of this kind, and others have produced similar results. But, the authors expand on this by looking at the ecology of the areas. These areas are affected by some of the layer marine heat waves on the record, and the authors take advantage of a large number of studies performed by other researchers (they determined that there are eight oceanic heat waves that can be studied in sufficient detail). They also focus their examination on three critical species: corals, seagrass and kelp. In general, ecosystems suffer under marine heat waves. The only things that clearly do alright are the ones that can move. Both fish and mobile invertebrates seem to manage the heat waves reasonably well; fish, in fact, saw an increase in diversity as tropical species began moving into unoccupied habitats. The authors suggest mobile species had two advantages; they can move to cooler waters if necessary and their ability to change habitats has left them more tolerant to a broad range of temperatures. The exception to this is birds, which are among the most severely affected group; they suffered because of changes in the availability of prey. Corals which bleach at high temperatures also did poorly.

The scientists then looked specifically at species that were near the edge of their temperature tolerance—where the heat waves hit the part of their habitat which was closest to the equator. Data from a set of 300 species showed that those were especially sensitive to marine heat waves; the authors also cite data showing that these heat waves can shift the equatorial limits of species by as much as 100 km. Finally, all these key species they focused on were hit in different ways by the oceanic heat waves. Corals tended to bleach, leaving them vulnerable to death; seagrass grew at lower densities; and the total biomass of kelp declined. These species tend to provide both habitats and resources for many others, hence, these problems can cascade across the ecosystem.

While these heat waves are taking place against a backdrop of general ocean warming, the dramatic effects seen here seem to be driven by the heat waves, which drive sudden, radical changes. While some of those may be reversed over time—assuming another heat wave does not come along—it is not clear whether all of them will or on what time scales. That, the scientists argue, means understanding average temperature is not good enough here. "The main focus of ecological research has been on trends in mean climate variables," they write, "yet, discrete extreme events are emerging as pivotal in shaping ecosystem, by driving sudden and dramatic shifts in ecological structure and functioning."

As per the United Nations, ocean heat hit a record high in 2018, raising urgent new concerns about the threat global warming is posing to marine life. In its latest State of the Climate Overview, the World Meteorological Organization (WMO) reaffirmed that the last four years had been the hottest on record—figures previously announced in provisional drafts of the flagship report. But, the final version of the report highlighted worrying developments in other climate indicators beyond surface temperature. "2018 saw new records for ocean heat content in the upper 700 m," a WMO statement said. 2017 also saw new heat records for the ocean's 2000 m, but, data for that range only goes back to 2005. The previous records for both ranges were set in 2017.

The United Nations Secretary-General Antonio Guterres described the latest findings as "another strong wake-up call" for governments, cities, and businesses to take action. The United Nations is hosting a major summit on September 23 that is billed as a last chance opportunity for leaders to tackle climate change, which Mr. Antonio Guterres has described as the defining issue of our time. The UN chief has urged world leaders to come to the summit with concrete plans to reduce GHGs emissions by 45% over the next decade and to net zero emission by 2050.

For a healthy Planet

In mid-March 2019, in Nairobi, world governments welcomed the Global Environment Outlook 6: Healthy Planet, Healthy People (GEO-6) report. GEO-6 argues that in a "business as usual" scenario, the world will exhaust its energy-related carbon budget in less than 20 years to keep the global temperature rise to well below 2 °C. It will take even less time than this to exhaust the budget to keep the global temperature rise below 1.5 °C.

GEO-6 shows that the interlocking environmental crises kill millions prematurely and affect and displace billions. Substituting for nature by buying air purifiers, building coastal defence systems to compensate for degrading mangroves, or just cleaning beaches is very expensive. Ironically, such costs increase the GDP as currently calculated. As Gross Domestic Product (GDP) grows at the cost of the environment and does not reflect an increase in everyone's well being, countries like India should reconsider how it calculates its Gross Domestic Product.

A country like India could save up to US $ 3.3–8.4 trillion in a 1.5 °C world. It is in India's interest to aim for 1.5–2.0 °C. This would mean investing in not new fossil fuels, but, in renewables and better batteries. Investing in inappropriate infrastructure has costs in terms of climate change and stranded assets, for instance, decommissioning oil and gas infrastructure in a small country like The Netherlands would cost anything between € 6.7—10 billion. If Indian universities develop tomorrow's technologies, it could provide cutting-edge and frugal technologies. This could change energy geopolitics and remove the excuse of rich countries of postponing carbon neutrality. Developing countries can change, for instance, Costa Rica, has pledged carbon neutrality by 2021.

A healthy planet is a public concern and good governments should take responsibility for its proper preservation and well being. When they hand responsibility to the private sector, clean air is only available to those who can pay for an air purifier. Poor people cannot afford air purifiers. Investing in water and sanitation will bring returns—a US $ 1 investment in water and sanitation could bring US $ 4 in returns; a green investment of 2% of global GDP could lead to similar growth rates by 2050. We must mobilize think tanks to work out context-specific solutions, be it in India, or in any other poor/developing country in the world facing this very grave environmental catastrophe, emanating from global warming.

Investing in education for sustainable development, vertical and compact cities, public transport with cheap parking facilities, renewable energy, removing single-use plastics, and reducing food waste, especially in a country like India, where enormous lunch and dinners are arranged during marriage functions, and, so much food is wasted, being simply thrown into dustbins, are the ways to reduce global warming.

Environmental costs of dietary habits:

Even dietary habits count when it concerns the hazards of global warming. According to the most recent data released by the Food and Agriculture Organization (FAO), Rome, on global food demand projections, the world needs to bridge a 70% "food gap" between crop calories available in 2006 and expected demand by 2050. This food gap stems primarily from the burgeoning global population growth in the developing world, India included, and, the changing diet pattern. For example, in India, with a current population around 1.3 billion, a large number are in prime youth whose food demand is large. A casual survey in many metros of the country reveals that food outlets are mushrooming like anything: hamburgers stuffed with beef and/or chicken (rarely pork or fish) are the order of the day for the young. Simply put, the young Indian, in a metro like Bengaluru, in southern India, is fast imitating his/her counterpart in US/Europe. *Idli, dosa, vada,* the once popular south Indian vegetarian fare have simply vanished from the food joints and given place to the new hamburgers. Meat is the predominant ingredient in these hamburgers.

The global population is projected to grow to 10 billion by 2050, with nearly two-thirds projected to live in cities. Additionally, at least 3 billion are expected to join the middle class by 2030, most of whom will be meat eaters (primarily beef). As nations urbanize and citizens become wealthier, people generally increase their calorie intake. Interestingly, this enhanced calorie intake comes mainly from meat and dairy products. China is the classic example of this change and India is following suit.

The dietary shift in India is unequivocally toward increasing flesh consumption. The Kentucky Fried Chicken (KFC) joints are more popular than the *masala dosa, puri bhaji* (the once popular vegetarian Indian fare) outlets. Together, these trends are driving a convergence toward western-style diets, with grilled beef steak or roasted pork or mutton, which contain high amounts of protein, leading to more calorific value. It is common to see many youngsters, including children, on the verge of obesity, aggravated by enhanced intake of sugary soft drinks. Although some of the shifts may result in better health and welfare gains for many people, the scale of convergence in diets will make it harder for the world to achieve several of the United Nations Millennium Goals, including those on hunger, healthy lives, better water management, combating global warming and climate change, which is the central theme of this book.

The crucial questions: A basic premise one has to understand is that far too many people, if one takes the example of countries like the United States, eat far too much than what is actually required for their normal daily metabolism. On average, a North American in USA or Canada, European or Russian consumes as much as 75–90 g of protein daily, of which, only about 30 g come from plant sources and more than 50 g come from animal sources; an adult with an average weight of 62 kg needs no more than 50 g of protein daily.

The World Resources Institute (WRI), located in the USA, India, China and Indonesia, estimates that the global demand for beef may increase by a whopping 95% by 2050. This is despite the fact that beef eating, especially eating red meat, as is the case with many Americans, has dropped in the US due to chronic heart diseases.

Breeding cattle for beef makes a huge demand on land for pastures, as nearly 25% of the land mass would be needed for pastures. Cattle rearing contributes to global warming from the cow belch, which contains methane, a potent greenhouse gas (GHG), that could contribute as much as 60% to global warming. The WRI specifically recommends, "reduce beef consumption." Cutting down on beef in daily diet offers both dietary and environmental benefits. The environmental benefits are very clear: it saves land for agriculture and reduces GHGs.

The Indian example: Many Indians are vegetarians. Why not promote vegetarianism globally to save the planet from the hazards of global warming? In terms of land use, it takes a lot more to produce a kg of beef compared to other food products. Take for instance the following example.

It is incredible how much resources are needed to produce a kg of beef: It requires 15455 l of water, 6.5 kg crop (grain), 330 m^2 of land, and releases 16.4 kg of CO_2. With 15455 l of water one can grow 60 kg of potatoes, 83 kg of tomato or 118 kg of carrots. On the African continent many children can be fed with 6.5 kg of grain. The cultivation of 1 kg tomato releases just 0.085 kg CO_2 compared 16.4 kg of CO_2 to produce 1 kg of beef. In other words, a kg of beef production releases into the atmosphere 19294% more CO_2 compared to the production of a kg of tomato, and CO_2 is an important GHG contributing to global warming. The impact of beef production, thus, contributes dramatically to global warming.

But, the central question is, why are most, if not all, westerners, Chinese, Brazilians, etc. love to eat beef? It is only because of the palatability. One would suggest, if beef is unavoidable, one can reduce its consumption to just a few times a week, not on a daily basis. In South Africa, the white population eat beef even for breakfast!

Another important aspect of the Indian lifestyle is, it was a circular economy, earlier, which is now fast changing and imitating the "use and discard" economy of the Western hemisphere. Single-use plastic bags are more popular now than cloth bags. So, why not cherish the re-users and recyclers? India had a judiciary that thought of long-term justice; why not protect that? Instead in India, one sees, instantaneous judicial decisions that can impair the "true spirit of justice". Justice relates to truth, fortitude to goodness, and *temperantia* to beauty; while prudence, in a sense, comprises all three said late Professor Schumacher, the author of the famous book "Small is Beautiful" One must debate, India included, where one must be in 2050, not just in 2020, and strive toward that, as the real goal.

A Sustainable Food System:

One of the Commissioners of the EAT-Lancet Commission said "The way we are producing food today is causing increased emission of greenhouse gases, depleting fresh water supply, compromising land use, exhausting the nitrogen and phosphorus cycle, and endangering biodiversity." The observation was made during the recent launch of the commission's report in New Delhi, India, in the first week of April 2019. He further added, "We need to find a safe space to provide food security to everyone by 2050." The report authored by 37 international experts has been put together by EAT, the science-based global platform for food system transformation, and the internationally reputed journal The Lancet.

The EAT-Lancet report, for the first time, proposes scientific targets for what constitutes a healthy diet derived from a sustainable food system.

With 1.35 billion people, that is 1 out of every 6 people, globally, India would soon surpass China to become the most populated nation in the world, and, that too on one-third of the land mass compared to that of China. Feeding all Indians a healthy diet in a sustainable manner, without compromising on the ecology and environment is going to be the most important challenge facing India, in the coming decades. If the food system is not fixed, it will be impossible to achieve the United Nation's Sustainable Development Goals. Thus, food, a healthy and sustainable way to eat, is the focal point, *vis-à-vis* global warming, which is threatening the very core of the food system.

Presenting some key steps required for the "great food transformation" Lawrence Haddad, Executive Director, Global Alliance for Improved Nutrition (GAIN), a global initiative launched by the United Nations in 2012, opined "Taxes on unhealthy foods, subsidies for healthier food options, strong leadership in the public and private sectors and strong civil society movements, is the need of the hour." In this context, it is a very healthy development that the young of the world is now a lot more vocal about the adverse impact of climate change, and teenagers in schools are taking the plunge into the "climate movement". One sees such movements in Sweden, Switzerland, etc.

It is the author's contention that bold leadership by the Western governments and Asian governments, in particular, Chinese, must come to grips with the menace of global warming. At the time of completing this book, it is heartening to note that Mr. Antonio Guterres, Secretary-General, United Nations, has called for a global summit on September 23, 2019, of all governmental heads, business leaders, scientists and policy makers, to discuss a future course of strategy to meet the crisis caused by global warming. One can only hope, it would not follow the pattern of the several "Food Summits" set up in Rome, by the Director General of the FAO, where global leaders met to discuss over lavish banquets the problem of hunger, while millions of Africans and Asians starved. The author of this book, boldly suggested during one of the "Food Summits" held in the early nineties "hunger should be discussed on empty stomachs," then only one would know the pangs of hunger. One ardently hopes, wishes and prays, the global summit on global warming does not follow suit!

Global warming and Dengue and Zika scare: Global warming could expose as many as one billion people, across the world, to the mosquito-borne diseases including Dengue and Zika virus by 2080, says a new study undertaken which examined temperature changes on a monthly basis worldwide. The study found that with the rise in temperature, Dengue is expected to have a year-round transmission in the tropics and several risks almost everywhere else. A greater intensity of infection is also predicted. To understand this, scientists at the Georgetown University in the US looked at temperature variation, month by month, to project the risks through 2050 and 2080.

While almost all of the world's population could be exposed at some point in the next 50 years, places like Europe, North America, and high elevations in the tropics that used to be too cold for the viruses will face new diseases like Dengue. On the other hand, in areas with the worst climate change—where steep increase in ambient temperature is encountered—including West Africa and South East Asia, serious reductions are expected for the *Aedes albopictus* mosquito, most noticeably in South East Asia, and West Africa, revealed the study, published in the scientific journal PLOS Neglected Tropical Diseases.

Both *Aedes aegypti* and *Aedes albopictus* mosquitoes can carry Dengue, Chikungunya and Zika viruses, as well as, at least a dozen other emerging diseases.

"Climate change is the largest and most comprehensive threat to the global health security," said Dr. Colin J. Carlson, Post Doctoral Researcher at the Georgetown University. "The risk of disease transmission is a serious problem, even over the next few decades" added Dr. Carlson.

Dengue is the fastest growing mosquito-borne disease across the world today, causing nearly 400 million infections every year, according to the World Health Organization (WHO).

The 2018 data from the National Vector Borne Disease Control Programme (NVB-DCP) and National Health Profile showed that cases of Dengue increased by 300%—from less than 60,000 cases in 2009 to 188, 401 cases in 2017.

1.1 Global Warming and Evolution of Wild Cereals

That climate variation, change, and stress are the major determinants of biodiversity and evolutionary change (Mathew 1939; Root et al. 2003) is true globally, regionally and locally (Nevo 2001). Among the array of environmental hazards that the world faces today, perhaps, none other is more serious than global warming. A recent report of an intergovernmental panel on climate change (IPCC 2014) provides an overview of this problem. Mercer and Perales (2010) have analyzed the effects of global warming on wildlife and crops, while Thornton and Cramer (2012) reviewed the effects of global warming on banana, barley, bean, cassava, chickpea, cowpea, pigeon pea, potato, faba bean, groundnut, lentil, maize, millet, rice, sorghum, millet, soybean, wheat, yam, forages, fisheries and aquaculture. Effects of global warming on oilseed rape pathogens has been reported by Siebold and von Tiedmann (2012), on rice by Liu et al. (2012). Global warming and its causes briefly discussed in the introductory chapter has become the most contentious issue of global, scientific, political and economic dimensions. There is, yet, no discernible discussion on the issue by some world powers, notably the US, but, there is an overall consensus that unless properly tackled, global warming could disastrously affect farming, as has been the case with the highly soil exploitative green revolution, and the consequences would be felt most on the poor.

1.1.1 Domestication: The Bedrock of Evolution

At the bottom of plant and animal evolution is their domestication, which, over millennia, led to the current human civilization. Darwin (1905) hypothesized that domestication is a gigantic evolutionary experiment of adaptation and speciation generating incipient species. Humans performed it primarily during the past 10,000 years (Zohary et al. 2012) mimicking speciation in nature. It leads to adaptive syndromes fitting human ecology (Harlan 1992). Domestication and the evolution of agricultural economies from preagricultural ones established human sedentary life, urbanization, culture, and, consequently an unprecedented scale of population explosion. The domesticated adaptive syndromes in cereals are non-shattering spikes and higher yields. Domestication makes the cultivars human-dependent, capable of surviving only under cultivation in human agricultural niches, to meet human needs and culture. In hindsight, perhaps, the most deleterious fallout of domestication was the overexploitation of fertile soils through unbridled chemical fertilizer use which led to the collapse of the so-called green revolution with all its attendant environmental hazards. Countries on the South Asian continent, in particular, India, and more specifically within the Indian States, the state of Punjab, is a tragic commentary of this collapse.

1.1.2 The Progenitors of Cultivated Barley and Wheat

Wild barley, *Hordeum spontaneum*, Wild emmer wheat *Triticum dicoccoides*
Zohary et al. (2012) have extensively reviewed domestication of plants in the Old World. Nevo (2012a, 2014a) reviews are primarily ecological genetics. Reviews of Nevo publications can be accessed at http://evolution.haifa.il, both regionally in the Near East Fertile Crescent (1975–2014), and locally at "Evolution Canyon", e.g.,

SSR divergence in wild barley (Nevo et al. 2005). For additional examples, see http://evolution.haifa.ac.il (1991–2013).

The relevance of global warming and its impact on agriculture is contextually very important to Israeli agriculture because of its geographic position. The desert climate of the country has not inhibited in its progress in agriculture. The common perception of Israel is that the Israelis have made the "desert to bloom". Hence, the following discussion would pertain to the Israeli experience, and discussing the use of CWR of barley and wheat in combating global warming.

1.1.3 Performance of Wild Cereals Juxtaposed to Three Decades of Global Warming

It is important to note that aridization increased in wheat growing areas, and in the region of origin and diversity of wild cereals in the Near East Fertile Crescent. Hence, it is of importance to follow the fate of wild progenitors, described earlier, and highlight the consequences for human food security. The projections for climate change in Israel is rather dire. The Ministry of Environment (2012) predicts an increase of average ambient temperature by 1.6–1.8 °C, specifically 1.5 °C before 2020 and 3.5 °C in 2071–2100, a 10% reduction in precipitation by 2020 and a 20% reduction by 2050, a 10% increase in evapotranspiration, delayed winter rains, increased rain intensity and shortened rainy seasons, greater seasonal variations in temperature, increased frequency and severity by extreme climate events, greater spatiotemporal climatic uncertainty and a tendency toward more arid climate. These predictions are in conformity with those of the Intergovernmental Panel on Climate Change Predictions, based on climate changes during 1960–1990.

The following table (Table 1.1) provides a list of drought-resistant populations of wild emmer wheat, *Triticum dicoccoides* (TD) and wild barley, *Hordeum spontaneum* (HS), with basic climatic characteristics, selected for drought tolerance in Israel.

1.1.4 Evolution of Wild Cereals During the Last Three Decades in Israel

It is crucial to evaluate the evolutionary adaptation of the wild progenitors of emmer wheat and barley with regard to global warming taking into consideration the importance of crop improvement. The researchers, consequently, examined ten populations of wild wheat, *Triticum dicoccoides* (hereafter TD), and ten populations of wild barley *Hordeum spontaneum* (hereafter HS) in Israel from 1980 to 2008, collected in the same sites in both years. The seeds of both collection years have been deposited in the gene bank of the Institute of Evolution, University of Haifa. These populations have been studied ecologically and genetically since 1975 by Nevo (ref: http://evolution.

Table 1.1 List of drought-resistant populations of wild emmer wheat (*Triticum dicoccoides*) and wild barley (*Hordeum spontaneum*) with basic climatic characteristics, selected for drought tolerance experiments in Israel

Wild emmer wheat			Wild barley		
Mean annual temperature (°C)	Mean annual rainfall (mm)		Mean annual temperature (°C)	Mean annual rainfall (mm)	
Population			*Population*		
Mt. Hermon	11	1400	Mt. Hermon	11	1400
Qazrin	18	530	Rosh Pinna	19	697
Rosh Pinna	18	697	Tabigha, terra rossa	24.1	436
Yehudiyya	19	550	Tabiga, basalt	24.1	436
Tabigha, terra rossa	24	436	Bet Shean	22.8	290
Tabigha, basalt	24	436	Mehola	23.0	290
Mt. Gilboa	21	400	Wadi Qilt	24.8	144
Kokhav Hashahar	20	400	Eizariya	20.0	380
Taiyiba	19	400	Talpiyyot	18.2	486
Sanhedriyya	17	548	Sede Boqer	19.4	91

Fig. 1.1 Wild emmer wheat, Triticum dicoccoides progenitor of all cultivated wheat

haifa.ac.il) and Nevo and Beiles (1989) (ref: Table 1.1). The results detailed in Table 1.1 and shown in Fig. 1.1 display distinctive adaptive phenotypic and genotypic changes which evolved in these wild cereals during a period of three decades of global warming in Israel, from 1980 to 2008, with regard to flowering time (FT) and simple sequence repeats (SSR) allelic turnover. These cause concerns as to the fate of these wild cereals and suggest the use of adaptive genetic changes which occurred in both TD and HS during this period, for future crop improvement.

1.1.5 Phenotypic and Genotypic SSR Results

Phenotypic results: The authors compared 20 natural populations of wild cereals ranging across 350 km in the country (at the Aaronsohn Agricultural Station's green house) which were subjected naturally to 28 years of global warming (Nevo et al. 2012). Comparison involved the time from germination until flowering of 800 geno-types under dry (300 mm) and wet (600 mm) irrigation regimes. Remarkably, in all TD and HS populations, without exception, the 2008 populations reached FT (Flow-ering Time) earlier than those collected in 1980 (significance test, $P < 10^{-15}$). All populations of HS and TD displayed, without exception, earliness (Fig. 1.1). The shortening of FT was highly significant in all 20 populations. Also, the shortening of FT was greater in HS than in TD population. The average shortening for each pop-ulation after the 28-year period in TD was 8.53 days (range 7.49–10.45), whereas in HS it was 10.94 days (range 8.21–17.26). The difference between species was significant (Wilcoxon rank sum test, $P < 0.01$, Nevo et al. 2012). Greenhouse plants under 600 mm irrigation flowered earlier than those under 300 mm.

Genotypic results: SSR marker analysis of about 15 individual samples for each population of the two wild cereals in both sampling periods (1980 and 2008) revealed remarkable genetic divergence, especially much more in TD population than in HS population in response to 28 years of climate change (Fig. 1.2). Allele depletion in 2008 compared with 1980 was found for both species of wild cereals. In TD, the total allelic count in 1980 was 318 alleles versus 290 alleles in 2008, a highly significant reduction of 28 alleles (8.%, $P < 0.0001$). Population allelic counts in 1980 were 113–173 alleles, whereas counts in 2008 were 104–157. The largest reduction was −46 alleles in the population of Sanhedriyya, near Jerusalem, which is an ecolog-ically marginal population located near the western border of the Judaean Desert. Seven populations showed reduction, and three showed increased allele count, sug-gesting the increase in new allelic count associated with global warming, though, this hypothesis needs further exploration. In HS, the total allelic count in 1980 was 319 and 309 in 2008, indicating a nearly significant ($P = 0.082$) reduction of 10 alleles and the same reduction trend as in TD, though lower in magnitude, possibly related to the higher hardiness of HS compared to TD. The population counts in 1980 were 94–144, whereas in 2008 it was 82–149 alleles. The largest reduction, −57 was in Mt. Hermon population, which represents an extreme ecologically marginal and cold steppic population. In TD, allele reduction was negatively correlated with alti-tude (−0.854*), humidity (−0.673*), and nearly significant with plant formation (−0.568*). In HS, without the Mt. Hermon extreme population, the difference was positively correlated with rainfall (0.790*), but, negatively correlated with evapora-tion (−0.692*) and plant formation (−0.867**). In other words, difference in allelic content increased in the more humid Mediterranean region, but, decreased toward the desert.

There are sharp genetic differences in allele frequencies within and between the two wild cereals. In all TD populations, there was at least one new allele that reached fixation and at least one fixed allele which was lost. By contrast, in HS, a new

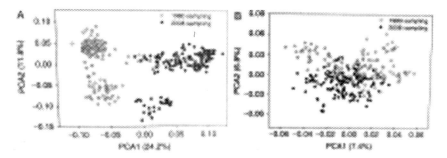

Fig. 1.2 Genetic associations of individual wild emmer wheat and wild barley plants. (Source: Nevo et al. 2012)

allele reached fixation in one population (Mehola, a steppic highly polymorphic population in the Jordan Valley, (near Bet Shean) only, and a fixed allele was lost in three populations (Rosh Pinna, Eizariya, and Mehola, the last two typical steppic populations). The average frequency of the new allele was 0.474 (0.363–0.805) in each TD population and 0.193 (0.151–0.244) in HS population. The difference between the two wild cereals was significant (P < 0.01, Wilcoxon rank sum test). Remarkably, the lost and the newly introduced alleles were widely distributed over the chromosomes, but, the lost alleles were smaller in size than the new ones for both species. Notably, the total variance between 1980 and 2008 was 20.4% in TD against only 4.4% in HS. Overall, the SSR response of TD to climate change was much stronger in the 28 years investigated and documented to undergo global warming more than HS, as expected from a relatively more ecologically specialized TD and hardier generalist HS, with a wider geographical range. These results are graphically illustrated in Fig. 1.2 as follows.

1.1.6   Effect of Global Warming on Phenotypic and Genotypic Turnover of Wild Emmer Wheat and Wild Barley in Israel

The afore-discussed report is the first one of its kind on the effect of global warming in Israel (Nevo et al 2012). Global warming is the only factor that could have triggered the earliness and allelic SSR turnover across the 20 natural populations (10 each of TD and HS) across 350 km from north Mediterranean Israel to its southern desert regions. In Israel, as in other hot climatic regions, the most important climatic factors are ambient temperature, rainfall, and evaporation, which are highly interrelated, with no possibility to disentangle them. The climatic predictions on global warming in Israel, provided by the Israeli Meteorological Survey, are based on climatic changes from 1960 to 1990 (Ministry of Environmental Protection 2012). Likewise, these predictions rely on facts of increased anthropogenic CO_2 emissions from 50 million tons in 1966 to 65 million tons in 2007—an increase of 30% in 42 years, which works to 0.7% increase per annum. The constant increase of CO_2, because of energy production, will greatly impact ambient temperature, rainfall, and evaporation. While no specific climatic records are available at each of the 20 populations of wild cereals investigated, the available evidence and predictions indicate rising temperature and declining rainfall over the past three decades. Moreover, the existing climate data in 1980 and 2008 imply that climate differences between the two years were representative of overall trends. Greater responses to climate change in the xeric compared to the mesic populations are expected and empirically confirmed (Nevo et al. 2012).

References

Darwin C (1905) The variation of animals and plants under domestication. John Murrayt, London, p 566. (popular edition in 2 volumes)

Harlan JR (1992) Crops and man, 2nd edn. American Society of Agronomy, Madison, WI

Liu L, Wang E, Zhu Y, Tang L (2012) Contrasting effects of warming and autonomous breeding on single-rice productivity in China. Agric Ecosyst Environ 149:20–29

Mathew WD (1939) Climate and evolution. New York Academy of Sciences, New York

Mercer KL, Perales HR (2010) Evolutionary response of landraces to climate change in centres of diversity. Evol Appl 3:480–493

Ministry of Environmental Protection (2012) Israel's Second National Communication on Climate Change (2010) Submitted Under the United Nation Framework Convention on Climate Change. www.environment.gov.il. Accessed 3 Nov 2014

Nevo E (2001) Evolution of genome-phenome diversity under environmental stress. Proc Natl Acad Sci USA 98:6233–6240

Nevo E (2012a) "Evolution canyon", a potential microscale monitor of global warming across life. Proc Natl Acad Sci USA 109(8):2960–2965

Nevo E, Beiles A (1989) Genetic diversity of wild emmer wheat in Israel and Turkey. Theor Appl Genet 77(3):421–455

Nevo E, Beharav A, Meyer RC (2005) Genomic microsatellite adaptive divergence of wild barley by microclimatic stress in "Evolution Canyon". Israel Biol J Linn Soc 84:205–224

Nevo E, Fu Y-B, Pavlicek T, Khalifa S, Tavasi M, Beiles A (2012) Evolution of wild cereals during 28 years of global warming in Israel. Proc Natl Acad Sci USA 109(9):3412–3415

Root TL, Price JT, Hall KR, Schneider SH, Resenzweig C, Pounds JA (2003) Fingerprints on global warming on wild animals and plants. Nature 421:57–60

Siebold M, von Tiedmann A (2012) Potential effects of global warming on oil seed rape pathogens in north Germany. Fungal Ecol 5:62–72

Thornton P, Cramer L (2012) Impacts of climate change on the agricultural and natural resources within the CGIAR's mandate. In: CCAFS working paper 23. CGIAR research program on climate change. Agriculture and food security (CCAFS), Copenhagen, Denmark. www.ccafs.cgiar.org. Titles in this working paper series aim to disseminate interim climate change, agriculture, and food security research and practices and stimulate feedback from the scientific community. Accessed 3 Nov 2014

Zohary D, Hopf M, Weiss E (2012) Domestication of plants in the old world, 4th edn. Oxford University Press, Oxford

Chapter 2
Wild Cereal Cultivation in Israel—Global Warming: An Important Link

That the Israeli climate has undergone a sea change during the last three decades is beyond dispute. However, the important point to ponder is how it has affected wild cereal growth and performance. The above-discussed results clearly show that all the twenty populations of the wild cereals, both emmer wheat and barley, displayed, without exception, similar tendencies of earliness and SSR allelic turnover, as a response to global warming. No other conceivable factor could result in essentially the same responsive patterns across species, populations, and ecologies. Pronounced earliness across Israel, in these cereals as a response to global warming is indicative of a general climatic effect operating both in mesic and xeric populations. Parmesan (2006) had theoretically predicted and practically demonstrated similar display of characteristics (ecological and evolutionary responses) paralleling the nonrandom response pattern demonstrated in other species, displaying similar ecological and evolutionary responses, and negating random walk. FT (Flowering Time) is selected in the major cereals. Flowering Time (FT)/Heading Date (HD) is an important evolutionary criterion for regional adaptation and yield in all cereals, with a clear genetic basis identified in both wild emmer wheat and wild barley as was observed in a garden experiment cultivation. The researchers grew 26 populations in HS and 12 populations in TD from across Israel in mesic (Mount Carmel) and northern Negev desert environments (Nevo et al. 1984a, b). The traits which were considered are performance in agronomically important phenotypic traits involved in 10 variables which compared and contrasted germination, earliness, and numerical and weight variables of biomass and final crop yield. Results in both wild cereals indicate that the characteristics investigated are partly genetically determined. Striking genetic variation was found within, but, more so between populations in each of the experimental sites, on Mount Carmel (mesic) and Northern Negev (xeric) regions. Remarkable environmental variation, including genetic environmental interaction, was found between the mesic and xeric sites, as well as between populations and years of investigation. The populations are not only varying in protein and DNA polymorphisms, but also, in quantitative traits of agronomic importance. These economically significant traits should be conserved and utilized in crop improvement. It will be important to

© Springer Nature Switzerland AG 2019
K. P. Nair, *Combating Global Warming*, Springer Climate,
https://doi.org/10.1007/978-3-030-23037-1_2

examine the effects of global warming on these traits, safeguard the germplasm collection, and subsequently use them in plant breeding efforts.

The experimental results of Nevo et al. (2012) indicate that Heading date/FT and yield are important criteria for regional adaptations and yield, as was recently reviewed in TD by Peng et al. (2012). Total earliness in both wild cereals across Israel is a red light that may predict the future extinction of these precious genetic resources (Nevo 2014). Global wheat production is already adversely affected by global warming with an estimated loss of 5.5% yield decline since 1980 (Lobell et al. 2011). Climate change impact on wild crop relatives shown in this report necessitates the continuous efforts for in situ and ex situ conservation of these important genetic resources for future crop improvement (Nevo 1998). The current global extinction of biodiversity exacerbated significantly by global warming also dictates coordinated conservation to protect the wild progenies from extinction. Conservation of these wild cereals needs to be both in situ and ex situ (Nevo 1998), to guarantee sustainable development and conserve their rich adaptive genetic diversity generally in the Near East Fertile Crescent and specifically in the small area of Israel in which both TD and HS are very rich in genetic resources for breeding, more than in countries 30-fold larger than Israel, like Turkey or Iran. For example, the total number of allozymic alleles detected in 27 loci in Israel, Turkey and Iran amounted to 127 (mean number of alleles per locus 4.7, range (1–17) decreasing as follows: Israel (103 alleles) greater than Turkey (63 alleles), and Iran (62 alleles) (Nevo et al. 1986). Note the similarity in alleles between Turkey and Iran, which are neighboring countries. The spatial patterns and environmental correlates and predictors of wild barley populations in the Near East Fertile Crescent are not only rich, but also partly adaptive and predictable by ecology, allozyme, and DNA markers. Consequently, conservation and utilization programs should optimize sampling strategies by following the ecological-genetic factors and allozyme markers as effectively predictive guidelines (Nevo 1987).

The SSR results are also equally informative as in the case of the Heading date/FT and yield discussed above. The general depletion of regulatory genetic diversity may cause deterioration of environmental adaptability. Remarkably, the SSR variance in response to global warming was clearly higher in the ecologically specialist TD than in the ecologically generalist HS (20% vs. 4% between years and samples respectively). Notably, specialist TD suffered more genetic reduction than generalist HS, which is more adapted to climate extremes and grows on varied soil types, whereas the reach of TD is restricted both climatically and edaphically. TD is more restricted geographically and does not grow in the true desert, in contrast to HS, which can grow on many soil types. TD grows primarily on basalt and terra rosa soil types (Nevo and Beiles 1989). Noteworthy however, and in contrast to the depletion of many alleles, some new alleles are detected in both wild species in 2008 (above reference), may be adaptive and valuable for crop improvement under global warming. Similarly, the genotypes and populations which displayed high yields under the stressful experimental regime (300 mm irrigation regime) may prove valuable in breeding for higher drought tolerance. The pattern of response to global warming in Israel may be relevant to the Near East Fertile Crescent, at large. The earliness of FT of wild cereals reported in the above-discussed findings is not an exception just to Israel, but, in all likelihood, applicable to the world, at large, with reference to the growth performance of current and future cultivars, growing under similar conditions of hot climate and dry environmental stresses. Both the wild cereals growing in the Near East Fertile Crescent, particularly in Northern Israel (Nevo et al. 2002, Saranga 2007) are rich in adaptive genetic resources against biotic and abiotic stresses, and are with high quality and quantity of storage proteins (glutenins, gliadins, and hordeins), amylases, and photosynthetic yields (Nevo 1989). Most of these resources, and many others yet to be discovered, are still untapped and are valuable for crop improvement (Nevo 2011).

Increasing risk from global warming will adversely affect food production, and hence, will be a threat to global food security. Inarguably, the cause for global warming forms the basis for this catastrophe. In the earlier section of this chapter, this author has discussed the role of excessive and unbridled use of chemical fertilizers, in particular nitrogenous, on the emissions of greenhouse gases, especially nitrous oxide, which constitutes up to 30% in the atmosphere and contributing substantially to global warming, because of its "long stay" in the stratosphere, for trapping the radiant heat. The ubiquitous green revolution, which has now fallen into disrepute, is the originator of this catastrophe. These aspects have been extensively discussed by Nair (1996, 2013) in prestigious publications, such as the Advances in Agronomy. While future crop improvement efforts should use both old and new adaptive genetic resources, because the current rich genetic maps of TD and HS permit the unraveling of the beneficial alleles of candidate genes and their introgression into cultivated wheat and barley, enhanced focus on intelligent soil management (Nair 2016) should also go forward, hand in hand.

2.1   Rice—Its Progenitors and Domestication

It is important to know that rice was domesticated in Asia and Africa separately. Quite probably, the Asian cultivated rice (*Oryza sativa*) must have been domesticated from the wild *Oryza rufipogon* during the last 1000 years. This variety (*Oryza sativa*) is the most dominant cultivated rice now globally. More recently, *Oryza glaberrima* (African rice) was domesticated probably from *Oryza barthii* in Africa. The Oryza genus (Vaughan et al. 2006) includes more than 20 species with the A genome clade of wild species (those most closely related to cultivated rice) including, in Asia, *Oryza rufipogon* and *Oryza nivara*; in Australia, *Oryza meridionalis*; and in Africa, *Oryza barthii* and *Oryza longistaminata*, and in South America, *Oryza glumaepatula*. Evolution of this group is not well understood. The tribe may be of Gondwanan origin, but, the current distribution may be a result of recent long-distance dispersal. The recent analysis of the poorly known populations in Northern Australia (Fig. 2.1) has revealed diversity which may contribute to a greater understanding of the evolution of the A genome species (Sotowa et al. 2013). More concerted efforts are needed to support both in situ and ex situ conservation of this key genetic resource for rice improvement.

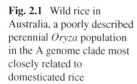

Fig. 2.1 Wild rice in Australia, a poorly described perennial *Oryza* population in the A genome clade most closely related to domesticated rice

2.1.1 Rice—Its Domestication Traits

For optimal agricultural productivity, one must target and make use of the domesticated plants, which have traits that make them suitable for agricultural production which often distinguishes them from the wild plants from which they are derived. Molecular analysis unravels the genetic basis of changes in plant architecture during domestication (Jin et al. 2008). A key domestication trait in the grasses domesticated cereals, is the resistance to shattering. Wild plants shed the seed at maturity to ensure dispersal. In agriculture, humans require a plant which retains seed so that it can all be harvested at once. Defining these domestication traits is important in understanding the constraints to the introduction of genes from wild plants into the domesticated gene pool. It has recently been suggested that the selection of nonshattering rice may have been preceded by the selection of rice with a closed panicle, resulting in the first step toward improved recovery of grain during harvesting (Ishi et al. 2013). Characterization of domestication-related genes in wild plant populations has been simplified by advances in genomics techniques (Malory et al. 2011).

2.1.2 Rice Domestication—Conflicts Between Natural Selection and Human Selection

In the domestication exercise, human selection may involve selection of traits desirable for human needs, but, not necessarily optimal for plant survival in the natural environment. In these cases, human intervention ensures crop varieties survive in cultivation that would be at a disadvantage relative to others growing under natural selection. The fragrant trait in aromatic rice is a good example of this type of trait. Humans find the aroma of fragrant rice very attractive. A major component of the aroma, 2-acetyl-1-pyrroline, is attractive to humans. The trait results from the loss of function of an aldehyde dehydrogenase (Bradbury et al. 2005). This recessive trait has been chosen more than once by humans as evidenced by the different alleles of this gene, now found in fragrant rice from different regions. One common allele is found in both *Basmati* (of Indian origin), and jasmine rice, suggesting the wide dispersal of this attractive trait. This gene encodes the enzyme responsible for a key step in proline metabolism which is upregulated in response to abiotic stress in plants. The mutation contributing to desirable aroma makes the plants less tolerant of stress (Fitzerald et al. 2010). In this way, the attractive trait ensures survival in the wild. Evolution under human and natural selection take very different paths. Better understanding of these differences and the genes involved is important in the development of crop varieties that are better able to cope with climate change.

References

Bradbury LMT, Fitzgerald TL, Henry RJ, Jin Q, Waters DLE (2005) The gene for fragrance in rice. Plant Biotechnol J 3:363–370

Fitzgerald TL, Waters DLE, Brooks LO, Henry RJ (2010) Fragrance in rice (Oryza sativa) is associated with reduced yield under salt treatment. Environ Exp Bot 68:292–300

Ishii T, Numaguchi K, Miura K (2013) OsLG1 regulates a closed panicle trait in domesticated rice. Nat Genet 45:462–465

Jinn J, Huang W, Ganouj P (2008) Genetic control of rice plant architecture under domestication. Nat Genet 40:1365–1369

Lobell DB, Schlenker W, Costa Roberts J (2011) Climate trends and global crop production since 1980. Science 333:616–620

Malory S, Shapter FM, Elphinstone FM, Chivers IH, Henry RJ (2011) Characterizing homologues of crop domestication genes in poorly described wild relatives by high-throughput sequencing of whole genomes. Plant Biotechnol J 9:1131–1140

Nair KPP (1996) The buffering power of plant nutrients and effects on availability. Adv Agron 57:237–287

Nair KPP (2013) The buffer power concept and its relevance in African and Asian soils. Adv Agron 121:416–516

Nair KPP (2016) The nutrient buffer power concept for sustainable agriculture. Notion Press, Chennai, p 434

Nevo E (1987) Plant genetic resources: prediction by isozyme markers and ecology. In: Rattazi MC, Scandalios JG, Whitt GS (eds), Izozymes: current topics in biological research. Agriculture, physiology and medicine, vol 16, pp 247–267

Nevo E (1989) Genetic resources of wild emmer wheat revisited: genetic evolution, conservation and tilization. In: Miller TE, Koebner RMD (eds) Proceedings seventh international wheat genetics symposium. Institute of Plant Science Research, Cambridge, pp 122–126

Nevo E (1998) Genetic diversity in wild cereals; regional and local studies and their bearing on conservation ex-situ and in-situ. Genet Resour Crop Evol 45:355–370

Nevo E (2011) Triticum. In: Kole C (ed) Wild crop relatives: genomics and breeding resources, cereals. (Chapter 10), pp 407–456. Springer, Heidelberg

Nevo E (2014) Evolution of wild barley at "Evolution Canyon". Adaptation, speciation, domestication and crop improvement. Isr. Israel J. Plant Sci. https://doi.org/10.1080/07929978.2014.940783. (ahead of Print 2014:1–11)

Nevo E, Beiles A (1989) Genetic diversity of wild emmer wheat in Israel and Turkey. Theor Appl Genet 77(3):421–455

Nevo E, Beiles A, Gutterman Y, Storch N, Kaplan D (1984a) Genetic resources of wild cereals in Israel and vicinity. I. Phenotypic variation within and between populations of wild wheat, Triticum dicoccoides. Euphytica 33 (3), 717–735

Nevo E, Beiles A, Gutterman Y, Storch N, Kaplan D (1984b) Genetic resources of wild cereals in Israel and vicinity II. Phenotypic variation within and between populations of wild barley, Hordeum spontaneum. Euphytica 33 (3), 737–756

Nevo E, Beiles A, Zohary D (1986) Genetic resources of wild barley in the near east: structure, evolution and application in breeding. Biol J Linn Soc 27:355–380

Nevo E, Korol AB, Beiles A, Fahima T (2002) Evolution of wild emmer and wheat improvement. population genetics, genetic resources, and genome organization of wheat's progenitor, Triticum dicoccoides, p 364. Springer, Heidelberg

Nevo E, Fu Y-B, Pavlicek T, Khalifa S, Tavasi M, Beiles A (2012) Evolution of wild cereals during 28 years of global warming in Israel. Proc Natl Acad Sci USA 109(9):3412–3415

Parmesan C (2006) Ecological and evolutionary responses to recent climate change. Ann Rev Ecol Evol Syst 37:637–669

Peng JH, Sun D, Nevo E (2010) Wild emmer wheat, Triticum dicoccoides, occupies a pivotal position in wheat domestication process. Aust J Crop Sci 5(9):1127–1143

Saranga Y (ed) (2007) A century of wheat research—from wild emmer discovery to genome analysis. Isr J Plant Sci 55, 3–4

Sotowa M, Ootsuka K, Kobayashi Y (2013) Molecular relationship between Australian annual wild rice. Oryza meridionalis, and two related perennial forms. Rice 6, 26

Vaughan DA, Ge S, Kaga A, Tomooka (2006) Phylogeny and biogeography of the genus Oryza. In: Hirano HY et al (ed) Rice biology in the genomics era, pp 218–234. Springer, Heidelberg

Chapter 3
Climate-Evolution—The Interrelationship

The ability of plants to adapt to different climatic conditions under natural selection can be found through an analysis of genetic variation among wild populations from different environments. This analysis suggests strategies for adaptation of plants for agricultural production in a changing climate. Genetic variation in a distant wild relative of the rice *Microlaena stipoides* has been investigated along a transect encountering contrasting environments (Shaper et al. 2012). The availability of rice genomic sequence resources makes this an attractive system to study. This analysis suggests that greater diversity may be found in populations in more stressed environments. More detailed analysis of these types of systems should provide important insights into strategies for crop adaptation to climate change. Domestication of new species which are better suited to the new climates is one option (Shapter et al 2013).

References

Shapter FM, Fitzgerald TL, Waters DLE, McDonald S, Chivers IH, Henry RJ (2012) Analysis of adaptive ribosomal gene diversity in wild plant populations from contrasting climatic environments. Plant Signal Behav 7:1–3

Shapter FM, Cross M, Ablett G (2013) High-throughput sequencing and mutagenesis to accelerate the domestication of Microlaena stipoides as a new food crop. PLoS ONE 8(12):e82641

© Springer Nature Switzerland AG 2019
K. P. Nair, *Combating Global Warming*, Springer Climate,
https://doi.org/10.1007/978-3-030-23037-1_3

Chapter 4
The Adaptation Range of Wild Crop Species to Fluctuations in Climate Change

The major staple food crops, such as rice, wheat, maize, potato, barley and sorghum, and grain legumes, oilseeds, and many vegetable crops are variously adapted to current climatic zones around the world. All of these crops differ in optimum temperatures for growth and harvestable yield, and, also in temperature limitations for critical growth phases, such as pre-anthesis grain filling in cereals. In addition to mean temperature rises, spikes in very high heat stress can be expected, which exceed those which previously occurred, with devastating consequences on food production (Lobell et al. 2008). In most cases, although some genetic variation for tolerance of reproductive heat stress has been identified in certain crops (Hall 2011), the domestic gene pools lack the genetic diversity to enable plant breeders to select for the required heat and drought stress tolerances. For most crops, a genetic bottleneck occurred with domestication during the past 11000 years, with selection being made for rare mutations to provide domestication traits, such as non-shattering, reduced seed dormancy, and a larger seed size, for the transition from hunter-gatherer to stable stationary, agricultural/farming societies. Often, the newly domesticated crop was largely isolated from the respective progenitor wild relatives, with monophyletic or biphyletic evolutionary origins and reproductively and/or spatially isolated grains, to be founded on a very restricted subsection of the genetic diversity in wild relatives. In a few cases, such as, for cross-pollinating sorghum, there has been continuous two-way introgression with crop wild relatives. However, crop wild relatives grow mainly on uncultivated locations and are often adapted to a much wider range of environments and associated climatic and edaphic stresses. The greater versatility in plant adaptation could provide the additional genetic diversity required to enable crop adaptation to climate change, in conjunction with such agronomic measures as conservation tillage, and inter/relay cropping for more efficient land utilization. There are several challenges in the transfer of desirable traits from wild relatives to crops, in particular, separation of close linked traits which favor survival in the wild rather than enhance realization of harvestable yield. This requires not only repeated backcrossing of the desirable genes from the wild into improved domestic genotypes, but, possibly, further genetic manipulations to exclude close linked undesirable genes. A wide range of traits for components of drought tolerance, heat stress, and food

© Springer Nature Switzerland AG 2019
K. P. Nair, *Combating Global Warming*, Springer Climate,
https://doi.org/10.1007/978-3-030-23037-1_4

quality have been identified in wild relatives of various crops, and many remain to be discovered.

The focus of this chapter would be the outlines of current distribution of major and significant minor crops and comparison of these with the ecogeographical range of wild relatives with a presumed genetic basis for such adaptations. A multidisciplinary approach would be required to achieve a follow-up to the "green revolution" of the 1960s for agriculture to achieve increased productivity in increasingly challenging environments in most parts of the world.

4.1   Strategies of Genetic Diversity

To meet the challenge of abiotic stresses which emerge with global climate changes, there are three main approaches for genetic adaptation of crops. Primarily, global warming is believed to be due to a high accumulation of carbon dioxide (CO_2) which has been bandied around as the main culprit, notwithstanding the fact that nearly 30% of global warming is contributed by nitrous oxide (N_2O) which has a fivefold higher potential to capture radiation heat and remain in the atmosphere for a much longer period, thus contributing to substantial global warming. The origin of N_2O is in the high-intensity soil extractive agriculture, where unbridled use of nitrogenous fertilizers like urea is the main culprit. The depletion of soil carbon exacerbates this devastating environmental fallout, as is vividly seen in Indian States like Punjab, which was known as the "cradle of green revolution", in South Asia where the hallmark of the green revolution was the unbridled use of chemical fertilizers, like urea, coupled with other pesticides. Extreme droughts and floods in the temperate and subtropical zones and unprecedented severity of spikes in local hot weather stresses are some of the features of this extreme climate change (Lobell et al. 2008). Potential genetic adjustments to meet these challenges are the following:

1. Locating expressions of tolerance to abiotic stresses in the domestic gene pools of crop to sustain the production of these crops in their current respective cropping zones
2. Introgression of stress tolerance traits from the respective wild germplasm, especially where this has a wider climatic/ecological distribution than for the corresponding domestic crops and
3. Changing the distribution of crops to match the altered local climates such as contraction of the acreage of maize production out of current marginal zones in Africa and India (Lobell et al. 2008) to be replaced by more drought-tolerant crops such as sorghum and both major and minor millets. An important fallout of the green revolution was the overemphasis on rice and wheat, where unbridled application of nitrogenous fertilizers led to global warming, as explained above, neglecting the local millets, like Finger millet (*ragi*) (*Eleusine coracana* L.) and *bajra* (*Pennisetum typhoides*), both very highly nutritious compared to rice or wheat. The former is a very hardy plant and is the nutritious staple of people in

rainfed areas of Karnataka State, in southern India. The latter is also nutritious and the staple of many parts of northern India. Additionally, both crops have high medicinal value, like controlling diabetes. Water requirement for both is much less compared to that of wheat or rice. Yet, it is the food habits and palatability which decide against the choice of these two crops, by most Indians. The poor Indians prefer these crops because of their hardiness, and low input needs (water and fertilizer).

The current discussion pertains to the first two points, diversity within the domestic gene pools and introgression of wild crop relatives.

Consequent to a complex global warming pattern, both minimum and maximum temperatures will shoot up, some regions will have reduced precipitation with wider variability and intermittent floods, and, where rainfall is reduced, there will most likely be more cloudless and still nights conducive to frost events (Redden et al. 2014). Loss of grain yield is the natural consequence of these climate aberrations, when such events coincide with the reproductive phase of the crop cycle.

References

Lobell DB, Burke MB, Tebaldi C, Mastrandrea MD, Falcon WP, Naylor RL (2008) Prioritizing climate change adaptation needs for food security in 2030. Science 319:607–610

Hall AE (2011) Breeding cowpea for future climates. In: Yadav SS, Redden RJ, Hatfield JL, Lotze-Campen H, Hall AE (eds) Crop adaptation to climate change. Wiley-Blackwell, Chichester, UK, pp 340–355 Chap 15

Redden RJ, Hatfield JL, Vara Prasad PV, Ebert AW, Yadav SS, O'Leary GJ (2014) Temperature, global climate change and food security. In: Franklin K, Wigge P (eds) Temperature and plant development. Wiley-Blackwell, Sussex, UK, pp 181–202, Chap 8

Chapter 5
Importance of Crop Wild Relatives

In a plant evolutionary time scale, running to millions of years, with a very constrained genetic bottleneck in the selection of rare mutations for domestication traits (reduced shattering, reduced seed dormancy, and increased seed size) and natural selection under a generally mild environment without frequent extremes, crop diversification during the past 10000 years, comparatively, has been relatively recent. Agriculture is generally not conducted in extreme environments of hot dry deserts unless irrigation is available, or under near-freezing temperatures combined with short growing seasons at either high or low altitudes. With climate change, cropping regions at medium to low altitudes are likely to contract in the temperate to tropical zones, but, coupled with over a 25% growth in population and overcrowding in megacities, greater productivity will be needed from a reduced cropping area to meet the ever-escalating food needs. The exception may be with underutilized potential crop lands at high altitudes. But, where most of the current world population lives, an expanded range of adaptation to climatic variables beyond what is available in the current domestic gene pools will require novel sources of genetic variation. The wider gene pools of the progenitor crop relatives are a natural first choice, as these are mainly still cross-compatible with the respective domestic gene pools and more easily utilized (Hancock 2012a, 2012b). Where this primary gene pool is inadequate to provide the necessary stress tolerances, it would be worthwhile to explore the secondary and tertiary crop relative gene pools, especially where their distribution is wide and encompasses extreme habitats (Snook et al. 2011). The wider genetic variance in most crop wild relative gene pools may retain greater potential than the domestic gene pools to adapt to wider extremes of temperature and drought.

References

Hancock JF (2012a) Cereal grains. In: Plant evolution and origin of species, 3rd edn. CABI, Wallingford, Oxfordshire, UK, pp 132–147, Chap 8

© Springer Nature Switzerland AG 2019
K. P. Nair, *Combating Global Warming*, Springer Climate,
https://doi.org/10.1007/978-3-030-23037-1_5

Hancock JF (2012b) The dynamics of plant domestication. In: Plant evolution and origin of species, 3rd edn. CABI, Wallingford, Oxfordshire, UK, pp 114–131, Chap 7

Snook LK, Ehsan Dullo M, Jarvis A, Scheldeman X, Kneller M (2011) Crop germ-plasm diversity: the role of genebank collections in facilitating adaptation to climate change. In: Yadav SS, Redden RJ, Hatfield JL, Lotze-Campen H, Hall AE (eds) Crop adaptation to climate change. Wiley-Blackwell, Chichester, UK, pp 495–506

Chapter 6
A Crop-Wise Comparison of Domestic Gene Pool with Wild Relatives on Ecogeographic Diversity

6.1 Wheat

Bread wheat, *Triticum aestivum*, is a hexaploid comprising an "A" genome from the wild diploid *Triticum urartu* (Au genome), a "B" genome most likely from *Aegilops speltoides*, and a "D" genome from *Triticum tauschii*. The first cultivated wheat was the diploid *Triticum monococcum*, einkorn wheat selected from wild *T.monococcum*, and both wild and cultivated *monococcums* are reproductively isolated from *T.urartu* with interspecific hybrids being sterile, although the two wild species have similar morphology (Nevo 2011). *T.urartu* has only limited distribution in the western Mediterranean, and mainly on basaltic soils and often in mixed stands with wild einkorn (Valkoun et al. 1998). Einkorn wheat was important in the Neolithic early agriculture, but now it is only occasionally grown in western Turkey, the Balkans, Switzerland, Germany, Spain, and the Caucasus (Nesbitt and Samuel 1996). Another wild diploid wheat is *T.boeticum*, the wild progenitor of *T.monococcum*, widely distributed in the southern Balkans and northern cool and wet regions of western Asia, including the Karakadag mountains where Einkorn may have been domesticated.

The emmer wheat is a tetraploid (AABB) genome, with the "A" genome derived from *T.urartu*, and the "B" genome from *A.speltoides* (Nevo 2011). The wild emmer (*Triticum dicoccoides*) is a subspecies of *Triticum turgidum* L., which includes several cultivated types of which the durum-free threshing wheat is the most prominent in agriculture. The *Triticum timopheevi* wheat is a tetraploid, but with an AAGG genome hybrids with *T.turgidum* L. are infertile. Wild emmer has restricted distribution, which is in the center of its diversity, which is Israel and Syria, and ranges from Jordan to Turkey and Iraq to Iran (Dvorak et al. 2011). Ecogeographically, it is associated with open stands in oak forests on basaltic-and limestone-derived terra rosa soils, ranging from altitudes 200 m below the sea level up to 1600 m above sea level in Israel and Iran, with mean temperature ranges of 11–24 °C and rainfall between 170 and 1400 mm spanning wild ecologies from semiarid to high rainfall conditions, and plant habits ranging from robust early maturing in low altitude arid areas to slender, late-maturing types at high altitudes (Nevo 2011). It is discontinuous

© Springer Nature Switzerland AG 2019
K. P. Nair, *Combating Global Warming*, Springer Climate,
https://doi.org/10.1007/978-3-030-23037-1_6

in "archipelago"—type population structures between Israel and Turkey, with local differentiation of polymorphism for allelic variation in both micro and macrogeographic scales (Nevo 2011). Allelic diversity in wild emmer is much wider than in the cultivated gene pool (Fu and Somers 2009). There is diversity for photosynthetic efficiency in *T.dicoccoides*, with populations from xeric-marginal regions being the most efficient (Nevo 2011). Diversifying and stabilizing selections enabled the species to adapt to multiple ecologic niches and stresses.

The evolution of the wild *T.dicoccoides* was very ancient, over 300,000 years (Dvorak and Ahkunov 2005). However, the evolution of the hexaploid bread wheat in the cultivated form was recent, from around 7000–8000 years, when the cultivated emmer wheat spread to the Caspian sea region to overlap the distribution of the wild *T.tauschii* (van Zeist and Baker-Heeres 1985). This evolution enabled the spread of wheat cultivation beyond Mediterranean climate to the cooler and more continental climate of Northern/Central Europe and Central Asia. The center of diversity of *T.tauschii* and ssp. strangulate is in the Caucasus, with the latter distributed only eastward to the Kopet Dag range near Iran, whereas *T.tauschii* ranges through Central Asia to China (Jones et al. 2013). The wide ecogeographic diversity of *T.tauschii* ssp. strangulata has contributed improved tolerance to drought stress in synthetic hexaploid wheat, reconstituted by combining tetraploid (mainly durum) and *T.tauschii*, in comparison with recurrent parents (Lage and Trrethown 2008), not as earlier maturity escapes, but, because of improved rates of grain filling and larger seed size, as evidenced by a comparison of derived synthetic hexaploids across a diverse range of *T.tauschii*, with diverse sources of bread wheat (Jones et al. 2013). Allelic diversity was much greater in *T.tauschii* than in the synthetic hexaploid wheat, which, in turn, had much greater diversity than the bread wheat. The latter grouped into a single exclusive subpopulation, the synthetics into three groups and *T.tauschii* accessions into seven groups, one of which from the Caspian sea region was considered to be the likely source of the D genome in bread wheat (Wang et al. 2013). This indicates that a much wider sampling of diversity in *T.tauschii* would add to the genetic variation for abiotic stress tolerance in bread wheat. Future analyses of the performance of synthetics derived from diverse sources of *T.tauschii*, across environments and variously introgressed into locally adapted and generally adapted wheat, could test the value of selecting the wild relative by ecogeographic source for specific stress tolerance.

6.2 Barley

Barley (*Hordeum vulgare*) is a diploid (n = 7) with *H.vulgare ssp.spontaneum* being the wild progenitor. These are infertile, although each is almost wholly self-pollinating (Morell and Clegg 2011) and post domestic introgression has continued to contribute to diversity in domestic barley (Russell et al. 2011). It is at low elevation that *H.vulgare* is mostly found and rarely above 1500 m as its cold tolerance is limited, with distribution from the Eastern Mediterranean to the Zagros mountains in

Western Iran (Zohary and Hopf 2000). The barbed lemmas of this wild barley attach well to animal fur, providing a dispersion mechanism. There are up to 30 additional wild *Hordeum* species of which only *Hordeum bulbosum* is cross-compatible with *H.vulgare* and is the only member of the secondary genepool (Morrell and Clegg 2011). *H.bulbosum* is perennial and outcrossing, with a wide geographic distribution in both the African and European regions of the Mediterranean and extending to Afghanistan and Tajikistan in Central Asia. The Western distribution is mostly of diploids (n = 7), while tetraploids predominate in Asia. This species contains valuable variation for abiotic stress tolerances, and for pest and disease tolerance (Johnston et al. 2009).

The large (5.5 Gb) genome of barley contains a large number of transposable elements. Three generations of backcrosses to the domestic gene pool are needed with introgression of wild alleles, to recover the domestic phenotype. The wild sources have contributed favorable alleles for grain yield and for yield component traits (Schmalenbach et al. 2009), as well as for disease resistance traits. The traits suitable for local adaptation of wild barley to various habitats include drought as well as frost tolerance, while the Mediterranean ecotype in comparison with desert and mountain types proved to be the most vigorous and competitive (Volis et al. 2004). The repeated presence of the nonbrittle rachis phenotype in separate loci in domestic barley is suggestive of multiple domestication, and similarly, for the six versus two row spike (Takahashi 1964; Komatsuda et al. 2007). This is supported by genetic associations between the land races and wild germplasm of the same region (Morgell and Clegg 2011). This also suggests further opportunities to explore in widening the domestic barley genepool to include wild relatives.

6.3 Rice

Oryza sativa L. and *Orayza glaberrima,* the domestic rice varieties, are diploid species (2n = 24), with 22 wild relative species, and together, comprise four species complexes: the cultivated rice complex and six primary gene pool relatives; the *officinalis* complex of nine species with five diploid and four tetraploid: the *meyeriana* complex with two diploid species; the *ridleyi* complex with two tetraploid species; and two other species of uncertain classification (Brar and Singh 2011). *Oryza sativa* is a major and important crop of Asia, especially southern, eastern, and South-East Asia, and also important in Southern Europe, like France, and the Americas, while *Orayza glaberirina* is distributed as a semi-upland (non irrigated largely, but, irrigated partially, as well) land type found in high rainfall areas. The hybrids of this with *Orayza sativa* are sterile. *O.sativa* has a diplylectic origin, and the two sub species, namely, *O.sativa* var. *indica* and *O.sativa* var *japonica* ecotype in temperate regions are clearly separated having diverged around seven million years ago with different ancestral genepools (Brar and Singh 2011). There is wide diversity within the *indica* group, and the wide diversity across the domestic *O.sativa* genepools may not be explained by suggested origins from *O.nivara* or from *O.rufipogon* (Sang and Ge 2007).

O.sativa are high yielders while *O.glaberrima* cultivars are not as high yielders as the former, but, carry many desirable traits as resistance to rice yellow mottle virus, African gall midge, and also to nematodes. It is also tolerant to drought, acidity and iron toxicity and strongly compete with weeds, which suggest opportunities for crossing of the two species for complimentary effects on desirable traits (Brar and Singh 2011). This strategy has produced the "Nerica" and other elite lines in breeding programs in West Africa and in the Philippines (Diagne et al. 2010).

The availability of wild relatives provides opportunities beyond the domestic genepool to adapt to increased levels of abiotic stresses expected with climate change. In the sativa complex, the seven primary genepool relatives, all diploids with the A genome as domestic rice, have diverse ecogeographic distribution, as follows: *Oryza nivara*, an annual, ranges from the Deccan Plateau in India to the Plateau regions of Myanmar, China and South-East Asia, sometimes cohabiting with *Oryza rufipogon*, a close relative. *O.rufipogon* is a perennial type found in tropical and subtropical Asia, South America and Australia. *Oryza longistaminata* occurring in Africa is closely related to the West African *Oryza barthii* sometimes in sympatric communities, which has a tall, (~2 m), erect, rhizomatous phenotype and is outcrossing. *Oryza breviligulata* is another wild relative found in Africa. *Oryza meridionalis* is found in the Australian tropics. And *Oryza glumaepatula* occurs in South America (Brar and Singh 2011). These wild relatives have undesirable traits, such as, shattering, poor grain type, poor plant type, low yield, and various incompatibility barriers, yet, useful traits have been transferred to domestic rice. Also, resistance to the deadly rice disease, bacterial blight, and to various other viruses, tolerance to soil toxicity, early maturing and cytoplasmic male sterility.

For domestic rice, the tertiary wild relatives are a further source of genetic diversity. *Oryza officinalis* complex with nine species—the B, C, D and E genomes variously in diploid and tetraploid combinations—have contributed resistance to the dreaded brown leaf hopper, bacterial blight disease, blast, and also to yield-enhancing traits from *Oryza grandiglumis* (CCDD) (Yoon et al. 2006). The wild F genome relative *Oryza brachyantha*, not classified within the four main species complexes, has provided resistance to bacterial blight and diversity for awning and for growth duration (Brar and Singh 2011). In the case of rice, introgression from wild relatives has provided improvement in both major gene and quantitatively inherited traits. There are polyphyletic origins among the cultivated rice species/sub species, and these also have the potential to add genes for improvement with the sharing of genepools. Yet, in the respective domestications, the dramatic transformation from a weedy type to reliable cultivars probably was associated with domestication gene bottlenecks. Given the wide range of habitats, ploidy levels, and plant types among the rice wild relatives, there is scope to increase their use in crop improvement, because of the encouraging results already obtained. It is unclear at present whether the diversity for abiotic stresses which will be needed for adaptation to climatic change can be obtained from wild relatives; however, these possibly may be found in the Australian ecologies, for the distribution of part of the primary genepool *Oryza rufigon*, and of the secondary genepool officinalis complex for *Oryza australiensis* and part of the *Oryza officinalis* distribution.

References

Brar DS, Singh K (2011) Oryza. In: Kole C (ed) Wild crop relatives: genomic and breeding resources, cereals. Springer, Heidelberg, Dordrecht, London, New York, pp 321–365

Diagne A, Midingoyi SKG, Wopereis M, Akintayo I (2010) The NERICA success story: development, achievements and lessons learned. The Africa Rice Center, Cotonou, Benin, p 29

Dvorak J, Ahkunov E (2005) Tempos of gene flow-deletions and duplications and their relationship to recombination rates during diploidy and polyploidy evolution in the Aegilops-Triticum alliance. Genetics 36:323–332

Dvorak J, Luo MC, Akhunov ED (2011) NI Vavilo's theory of centres of diversity in the light of current understanding of wheat diversity, domestication and evolution. Czech J Genet Plant Breed 47:S20–S27

Fu YB, Somers D (2009) Genome wide reduction of genetic diversity in wheat breeding. Crop Sci 49:161–168

Johnston PA, Timmerman-Vaughan GM, Farnden KJ, Pickering R (2009) Marker development and characterization of Hordeum bulbosum introgression lines: a resource for barley improvement. Theor Appl Genet 118:1429–1437

Jones H, Gosman N, Horsnell R (2013) Strategy for exploiting germplasm using genetic, morphological and environmental diversity: the Aegilops tauschii Coss. Example Theor Appl Genet 126:1793–1808. https://doi.org/10.1007/s00122-013-2093-x

Komatsuda T, Pourkheirandish M, He C (2007) Six row barley originated from a mutation in a homeodomain zipper I-class homebox gene. Proc Natl Acad Sci USA

Lage J, Trethowan RM (2008) CIMMYT's use of synthetic hexaploid wheat in breeding for adaptation to rainfed environments globally. Aust J Agr Res 59:461–469

Morell PL, Clegg MT (2011) Hordeum. In: Wild crop relatives: genomics and breeding resources, cereals, pp 309–319, Chap 6

Nesbitt N, Samuel D (1996) From staple crop to extinction? The archeology and history of hulled wheats. In: Padulosi S, Hammer K, Heller J (eds) hulled wheats. International Plant Genetic Resources Institute, Rome, Italy, pp 41–100

Nevo E (2011) Triticum. In: Kole C (ed) Wild crop relatives: genomics and breeding resources. Cereals. Springer, Berlin Heidelberg, pp 407–456, Chap 10

Russell J, Dawson IK, Flavell AJ (2011) Analysis of 1000 single nucleotide polymorphisms in geographically matched samples of landrace and wild barley indicates secondary contact and chromosome-level differences in diversity around domestication genes. New Phytol 191:564–578

Sang T, Ge S (2007) The puzzle of rice domestication. J Integr Plant Biol 49:760–768

Schmalenbach I, Leon J, Pillen K (2009) Identification and verification of QTLs for agronomic traits using wild barley introgression lines. Theor Appl Genet 118:483–497

Takahashi R (1964) Linkage study of two complimentary genes for brittle rachis in barley. In: Berichte des Ohara Instituts fuer Landwirtschaftliche Biologie, vol 38. Okayama Universitaet, pp 81–90

Valkoun J, Waines JG, Konopka J (1998) Current geographical distribution and habitat of wild wheats and barley. In: Damania AB, Valkoun J, Wilcox G, Qualset CO (eds) Origins of agriculture and crop domestication. ICARDA, Aleppo, Syria, pp 293–299

van Zeist W, Baker-Heeres JAH (1985) Archaeological studies in the Levant 1. Neolithic sites in the Damascus basin: Aswad, Ghoraife, Ramad. Paleohistoria 24:165–256

Volis S, Verhoeven KJ, Mendlinger S, Ward D (2004) Phenotypic selection and regulation of repro-
    duction in different environments in wild barley. J Evol Biol 17:1121–1131

Wang JR, Luo MC, Chen ZX, You FM, Wei YM, Zheng YL, Dvorak J (2013) Aegilops tauschii
    single nucleotide polymorphisms shed light on the origins of wheat D-genome genetic diversity
    and pinpoint the geographic origin of hexa-ploid wheat. New Phytol 198:925–937

Yoon DB, Kang KH, Kim KJ (2006) Mapping quantitative trait loci for yield components and
    morphologic traits in an advanced population between O. grandiglumis and the O. sativa japonica
    cultivar Hwaseongbyeo. Theor Appl Genet 112:1052–1062

Zohary D, Hopf M (2000) Domestication of plants in the old world, 3rd edn. Oxford University
    Press, New York, p 316

Chapter 7
Relevance of Wild Relatives in Other Crops in Plant Breeding Programs

When comparisons are made between domestic crops and their wild relatives, it must be recognized that both are from different origins of distribution, which vary by crop, and specific consideration of the primary, secondary and tertiary gene pools. In most cases crops have a wider ecogeographic distribution than their corresponding wild relatives, with manual selection for wider adaptation, continuing ever since domestication. However, wild relatives have had to persist without cultivation, weed management, and often under restrictive habitat to ecologies and environments unsuitable to domesticated farming. To varying extents, wild relatives can provide genetic diversity for adaptation to extreme environments and to hostile soils, and, in most cases offer much wider genetic diversity than in the case of domestic gene pool (Hancock 2012b). Many crops have been domesticated, but, remain largely confined to the geographic center of their origin, where domesticated. *O.globerrima* in West Africa; tef in Ethiopia; a large number of root, grain, and fruit species sometimes with specific adaptations to different altitudes, in the Andes, and many examples of underutilized crops worldwide (Padulosi et al. 2011; Hancock 2012b). Some crops, such as wheat, barley, and pea have two or more centers of diversity, and some do not have a clear center of diversity with polyphyletic origins such as sorghum and *Phaseolus* beans, and for sorghum as an outcrossing species, there has been continuous two-way introgression between the domestic and wild genepools, where their distribution overlap (Hancock 2012b). With chickpea, the distribution of the primary and secondary genepools are restricted to Turkey, and further genetic bottlenecks occurred post domestication with the conversion of domestic chickpea from a winter to a spring habit, thus avoiding the devastating winter pathogen *Aschochyta*. However, the tertiary gene pools of chickpea have wide ecogeographic distribution and include both annual and perennial types. For lentil and pea, the primary gene pools are widely distributed, as are the secondary and tertiary relatives of lentil, though, not for pea with the secondary gene pool in Ethiopia and the Middle East, and the tertiary gene pool (*P.vavilovia*) confined to high elevations in the Caucasus. Thus, the wild relatives differ as per their respective crops in their diversity of ecographical range and the potential tolerances of climate and soil stresses that they may offer to the domestic gene pool. These provide opportunities to be explored beyond major

© Springer Nature Switzerland AG 2019
K. P. Nair, *Combating Global Warming*, Springer Climate,
https://doi.org/10.1007/978-3-030-23037-1_7

gene traits for extending the adaptation range of crops to the extremes expected with climate change. Also, we need to learn more about which crops can benefit from the inclusion of the wild relative genepools. Such conjecture is too difficult to predict, and the success among different crops is likely to be very uneven.

References

Hancock JF (2012b). The dynamics of plant domestication. In: Plant evolution and origin of species, 3rd edn. CABI, Wallingford, Oxfordshire, UK, pp 114–131, Chap 7
Padulosi S, Heywood V, Hunter D, Jarvis A (2011) Underutilised species and climate change: current status and outlook. In: Yadav SS, Redden RJ, Hatfield JL, Lotze-Campen H, Hall AE (eds) Crop adaptation to climate change. Wiley-Blackwell, Chichester, UK, pp 507–521

Chapter 8
Conservation Research and Crop Wild Relatives Use

8.1 How Does Climate Change Affect Crop Wild Relatives (CWR)?

Climate change is one of the major threats for CWR species, as CWR fail to adapt to new climate conditions of their habitats (Jarvis et al. 2008). Both CWR taxonomic and genetic diversity will most likely be threatened by climate change, because of their common reliance on disturbed habitats and the lack of resilience of these habitats. At the same time, however, CWR also have an important role to play in climate change adaptation because they contain the breadth of genetic diversity necessary to combat climate change. An evaluation of possible threats posed by climate change of eleven wild gene pools of major crops, worldwide, comprising a total of some 343 species, using Intergovernmental Panel on Climate Change (IPCC) gas emissions scenario A2a of 18 global climate models for the year 2050 to map current and future predicted richness in CWR and the predicted change in richness was conducted by Jarvis et al. (2008) The map shows hotspots of change where significant loss of diversity is expected to occur mostly in Sub-Saharan Africa, eastern Turkey, the Mediterranean region, and parts of Mexico. Another investigation by Lira et al. (2009) in Mexico used bioclimatic modeling and two possible scenarios of climate change to analyze the distribution patterns of eight wild cucurbits closely related to cultivated species. The results showed that all eight taxa displayed a marked contraction in area under both climate scenarios, and, that under a drastic climatic change scenario, the eight taxa would be maintained only in 29 of 69 protected areas in which they currently occur. Similarly, other climate change impact studies on maize diversity in Mexico (Mercer et al. 2012) have shown suitable habitat of maize wild relative, *Tripsacum* spp. and *Zea* spp. to be significantly reduced, thereby increasing the risk of genetic erosion. As mentioned above, CWR can also be a solution for climate change adaptation.

One of the biggest challenges facing agriculture is how to cope with the impact of climate change on crop production, with rising temperatures and reduced precipitation in some parts of the world, as discussed in the early part of this chapter, for

© Springer Nature Switzerland AG 2019
K. P. Nair, *Combating Global Warming*, Springer Climate,
https://doi.org/10.1007/978-3-030-23037-1_8

example in Israel, discussed at length, and higher precipitation in others. Generally, rising temperatures and reduced precipitation will affect semiarid regions and reduce yields of maize, wheat, and rice over the next two decades (Brown and Funk 2008). It has been suggested that breeding of better-adapted crop varieties would be one solution to cope with the impact of climate change (Lobell et al. 2008). However, such a strategy would most likely be costly and would need new sources of genes. CWR have been acknowledged as an important future source of novel genes and adaptive traits, and, with the advances made in biotechnology, the transfer of sets of genes or even complexes of genes conferring tolerance to abiotic stresses, such as, drought, salinity, and temperature from more distantly related CWR into breeding programs is possible (Ford-Lloyd et al. 2011).

References

Brown ME, Funk CC (2008) Food security under climate change. Science 319:580–581

Ford-Lloyd BV, Schmidt M, Armstrong SJ (2011) Crop wild relatives—undervalued, underutilized and under threat? Bioscience 61:559–565. 250 Kodoth Prabhakaran Nair

Jarvis A, Lane A, Hijmans R (2008) The effect of climate change on crop wild relatives. Agr Ecosyst Environ 126:13–33

Lira R, Tellez O, Davila P (2009) The effects of climate change on the geographic distribution of Mexican wild relatives of domesticated Cucurbitaceae. Genet Resour Crop Evol 56:691–703

Lobell DB, Burke MB, Tebaldi C, Mastrandrea MD, Falcon WP, Naylor RL (2008) Prioritizing climate change adaptation needs for food security in 2030. Science 319:607–610

Mercer KL, Perales HR, Wainwright JD (2012) Climate change and the transgenic adaptation strategy: smallholder livelihoods, climate justice, and maize landracesexico. Glob Environ Chang 22:495–504

Chapter 9
The Threats to Crop Wild Relatives

In addition to climate change, there are several other threats to CWRs, such as, their loss, degradation, and fragmentation of their natural habitats, deforestation, logging, plantation and industrialized agriculture, forestry, dryland destruction and desertification, fire, urbanization, mining and quarrying, and, massive invasive species. CWRs are often associated with disturbed and preclimax communities, which are the same habitats, most subject to increasing levels of anthropogenic change and destruction (Jain 1975), and, these habitats are not being adequately conserved by ecosystem conservation agencies. Genetic erosion caused by the disturbance of natural habitat by the various human activities is a serious threat to CWR. For instance, documentation of the reduction of distribution of two wild rice species *Oryza rufipogon* Griff. and *Oryza officinalis* Wall. ex G. Watt in Guiping, Guanzi, China was carried out (ShiChun et al. 2007). The reasons that led to serious destruction of wild rice resources were land reclamation for agriculture, over herding, introduction of exotic species, and coal mining. The reduction of natural populations of *Oryza rufipogon* was also reported in the central plain of Thailand (Akimoto et al. 1999).

A further threat to CWR is the onslaught of diseases. It has been documented by Van Vianen et al. (2013). *Lepidium oleraceum* Sparrm.ex G. Forst is a threatened endemic species of New Zealand, because of habitat degradation, the loss of associated seabird and seal colonies, insect and fungal pests, and browsing by introduced mammals. The discovery of Turnip mosaic virus (TuMV; Family: Potyviridae; Genus *Potyvirus*) in glasshouse specimens and in a replanted individual of *L.oleraceum* from Banks Peninsula, Canterbury, New Zealand, were the first records of any virus infecting this species (Fletcher et al. 2009). This highlighted a potential new threat to *L.oleraceum*. TuMV can cause symptoms of leaf mosaic, necrosis, chlorotic mottle, and, severe distortion as well as reductions in plant biomass in important crop brassicas (Spence et al. 2007). Additionally, TuMV has been known to reduce the fitness of wild brassicas by reducing biomass, fecundity, and survival (Maskell et al. 1999).

© Springer Nature Switzerland AG 2019
K. P. Nair, *Combating Global Warming*, Springer Climate,
https://doi.org/10.1007/978-3-030-23037-1_9

9.1 CWR—Is There a Red List?

The fact that CWRs are threatened is real. The question, however, to what extent? Although there are 19,738 plant species on the International Union for Conservation of Nature (IUCN) Red List (IUCN 2014. IUCN Red List Version 2014.3 Table 3b), the number of CWR within these are not fully documented. In Europe, 161 species and 23 subspecific Euro-Mediterranean CWR taxa were included in the 2014 IUCN Red List of Threatened Species (most of these taxa being trees) (Kell et al. 2008). Subsequently, an extensive IUCN Red List assessment of high-priority European CWR species was carried out (Kell et al. 2012). In total, 571 native European CWR of high-priority human and animal food crop species were assessed: 313 (55%) were assessed as Least Concern, 166 (29%) as Data Deficient, 26 (5%) as Near Threatened, 22 (4%) as Vulnerable, 25(4%) as Endangered, and 19 (3%) as Critically Endangered. These assessments have been published in the first European Red List and those that are endemic to Europe (188 species) have been published in the IUCN Red List of Threatened Species. Out of the 591 CWR species for which regional assessments were carried out, 19 were assessed as Not Applicable (NA), and one species, *Allium jubatum* J.F. Macbr, was assessed as Regionally Extinct (RE). This investigation showed that the highest numbers of red-listed species are found in the countries of Southern and Eastern Europe, which are known to have large floras, and thus, large number of CWR species. In the United Kingdom, 13 CWR taxa have been assessed as threatened (Maxted and Kell 2009). In the USA, Khoury et al. (2013) mentions that 16 CWR taxa from the USA were assessed as extinct, endangered, or vulnerable (IUCN 2012), according to the IUCN Red List of Threatened Species. Sixty two taxa are listed as endangered under the US Endangered Species Act (Endangered Species Act of 1973, 16 U.S.C. Sec 1531), 10 taxa as threatened, and 11 taxa as candidates for listing (NatureServe 2009).

Under the United Nations Environment Programme (UNEP)/Global Environment Facility, GEF) supported project titled "In situ conservation of crop wild relatives through enhanced information management and field application", led by Biodiversity International, Red List Assessment of CWR was carried out for priority crops from five countries—Armenia, Bolivia, Madagascar, Sri Lanka, and Uzbekistan. In Bolivia, 152 species were assessed and seven species were found to be Critically Endangered, 22 Endangered, 16 Vulnerable, and 20 Not Threatened, while 62 species were of Least Concern, and 25 Data Deficient (VMABCC-BIODIVERSITY, 2009). The CWR portal (www.cropwildrelativesportal.org) provides more detailed information about the status of CWR from many countries.

References

Akimoto M, Shimamoto Y, Morishima H (1999) The extinction of genetic resources of Asian wild rice, *Oryza rufipogon* Griff: a case study in Thailand. Genet Resour Crop Evol 46:419–425

Fletcher JD, Bulman S, Fletcher PJ, Houliston GJ (2009) First record of Turnip mosaic virus in Cook's scurvy grass (*Lepidium oleraceum* agg) an endangered native plant in New Zealand. Aust Plant Dis Notes 4:9–11

IUCN (2012) The IUCN red list of threatened species. Vision 2012.I. International union for conservation of nature and natural

IUCN (2014) The IUCN red list of threatened species 14.3 (Table 3 updated November 2014). Available on http://www.iucnredlist.org/about/summary-statistics#Tables-3_4. Last reviewed 24 Dec 2014

Jain SK (1975) Genetic reserves. In: Frankel OH, Hawkes JG (eds) Crop genetic resources for today and tomorrow. Cambridge University Press, Cambridge, pp 379–396

Kell SP, Knupffer H, Jury SL, Ford-Lloyd BV, Maxted N (2008) Crops and wild relatives of the Euro-mediterranean region: making and using a conservation catalogue. In: Maxted N, Ford-Llyod BV, Kell SP, Iriondo J, Dulloo E, Turak J (eds) Crop wild relative conservation and use. CABI Publishing, Wallingford, UK

Kell SP, Maxted N, Bilz M (2012) European crop wild relative threat assessment: knowledge gained and lessons learnt. In: Maxted N, Dulloo ME, Ford-Lloyd BV, Frese L, Iriondo J, Pinheiro de Caravalho MAA (eds) Agrobiodiversity conservation: securing the diversity of crop wild relatives and landraces. CAB International Publishing, Wallingford, pp 218–242

Khoury CK, Greene S, Wiersema J, Maxted N, Jarvis A, Struik PC (2013) An inventory of crop wild relatives of the United States. Crop Sci 53:1496–1508

Maskell LC, Raybould AF, Cooper JI, Edwards ML, Gray AJ (1999) Effects of turnip mosaic virus and turnip yellow mosaic virus on the survival, growth, and reproduction on wild cabbage (*Brassica oleracea*). Ann Appl Biol 135:401–407

Maxted N, Kell SP (2009) Establishment of a global network for the in situ conservation of crop wild relatives: status and needs. Italy, FAO commission on genetic resources for food and agriculture, Rome, p 266

NatureServe (2009) NatureServe explorer: an online encyclopedia of life. Version 7.1

ShiChun L, ChengBin C, QingWen Y (2007) Investigation and protection countermeasures for wild rice resources in Guiping, Guangxi. Southwest China J Agric Sci 20:943–947

Spence NJ, Phiri NA, Hughes SL (2007) Economic impact of Turnip mosaic virus, Cauliflower mosaic virus and Beet mosaic virus in three Kenyan vegetables. Plant Pathol 56:317–323

Van Vianen JCCM, Houliston GJ, Fletcher JD, Heenan PB, Chapman HM (2013) New threats to endangered Cook's scurvy grass (*Lepidium oleraceum*. Brassicaceae): introduced crop viruses and the extent of their spread. Aust J Bot 61:161–166

Chapter 10
Gene Flow Between Cultivated Plants and Their Wild Relatives

More often than not, farmers, if they are in the know of things, tolerate the presence of CWR on the farms because they recognize their value in providing beneficial traits to their standing crops. Genes from wild plants have provided neighboring cultivated plants with resistance against pests and diseases and improved tolerance to abiotic stresses, such as salinity, drought etc., through natural crossing. Here one has to make a very important distinction between what can happen in the case of gene flow from crop wild relatives, and, what can happen when genetically modified crops are grown in the field. The gene flow here can have very deleterious effects like the evolution of super weeds as has been observed in the UK. The beneficial effects of CWR in providing useful traits to their domesticated crops can be demonstrated in home gardens, which provide gene flow between plant populations inside and out of the garden (Hughes et al. 2007). Gene flow involving CWRs and cultivated plants is indeed facilitated by the limited spatial separation of individuals grown in home gardens (Galluzzi et al. 2010). Because cultivated species have generally suffered strong bottlenecks through domestication (Doebley et al. 2006), gene flow involving wild species and their domesticated counterparts is valuable in the enrichment of their effective population sizes (Mercer and Perales, 2010). This geneflow often results in significant intraspecific diversity (Eyzaguirre and Linares, 2014), which not only increases a species chance to adapt and survive, over time, (Nunney and Campbell, 1993), but, also, provides crucial material for breeding (Feuillet et al. 2008) and for establishing, complementing, or restoring germplasm collections (Castifieiras et al. 2007). In Benin, an investigation of 240 home gardens reveals a total of 285 species being cultivated, of which 20 CWRs were identified, namely, *Amaranthus spinosus.* L., *C.colosynthis* (L.) Schrad, *Blighia unijugata* Baker, *Corchorus tridens* L., *C.trilocularis* L., *Dioscorea abyssinica* Hochst, ex. Kunth, *D.cayenensis* Lam., *Gossypium arboreum* L., *Ipomoea aquatica* Forssk., *I.eriocarpa* R. Br., *I.involucrata* P. Beauv. *M.glaziovii* Mull. Arg, *Ocimum americanum* L., *Pennisetum purpureum* Schumach., *Sesamum radiatum* Schum et Thonn., *Solanum erianthum* D. Don., *Solanum torvum* Sw., *Talinum triangulare* (Jacq.) Willd. *Terminalia glaucescens* Planch. ex. Benth., *Vernonia colorata* (Willd.) Drake. Gene flow is one possible avenue for surviving major shifts in biotic and abiotic conditions. It may be the key

© Springer Nature Switzerland AG 2019
K. P. Nair, *Combating Global Warming*, Springer Climate,
https://doi.org/10.1007/978-3-030-23037-1_10

Table 10.1 Some important examples of high gene flow between cultivated crops and their crop wild relatives

Crop	Remarks	Country	Reference
Groundnut (*Arachis hypogea*)	High gene flow likelihood between *A.hypogea* and its cross-compatible close relative *A.monticola*	Argentina	Anderson and de Vicente (2010)
Oats (*Avena sativa*)	High gene flow likelihood between *A.sativa* and its cross-compatible weedy relatives, *A.fatua*, *A.ludoviciana*, *A.occidentalis* and *A.sterilis*	North-East and South Africa, N and S America, Central Asia, Australia, Europe, United Kingdom and Middle East	Anderson and de Vicente (2010)
Mustard (*Brassica napus*)	Very high gene flow likelihood between *B.napus* and wild *B.rapa*	U.S.A, Europe, Africa, Asia and Australia	Anderson and de Vicente (2010)
Finger millet (*Eleucine coracana*)	High gene flow likelihood between *E.coracana*, and its wild ancestor *E.coracana* subsp. *africana* and probably for its tetraploid wild relative *E.kigeziensis*	Africa	Anderson and de Vicente (2010)
Soybean (*Glycine max*)	High gene flow likelihood between *Glycine max* and the annual wild soybeans belonging to the *Glycine soja* species complex	East Asia	Anderson and de Vicente (2010)
Potato (*Solanum tuberosum*)	High gene flow likelihood between *Solanum* and its wild relatives	Mountainous area from Mexico to Argentina	Anderson and de Vicente (2010)
Cotton (*Gossypium hirsutum*)	High geneflow likelihood between *Gossypium hirsutum* and its wild relatives *G.barbadense*, *G.tomentosum*, *G.darwinii*, and *G.mustelinum*	U.S.A	Anderson and de Vicente (2010)
Cowpea (*Vigna unguiculata*)	Extensive gene flow between wild (*Vigna unguiculata* ssp. *unguiculata* var. *unguiculata*; *V.unguiculata* subsp. *pubescens*, *V.unguiculata* subsp. *tenuis*, and *V.unguiculata* subsp. *alba*) and domesticated types of cowpea, confirmed by AFLP analysis. The investigation showed that var. *spontanea* originated in East Africa and spread eastward and southward. These data also showed that gene flow appears to have had a considerable impact on the organization of genetic diversity within *V.unguiculata*, resulting in a large number of weedy forms	Africa	Coulibaly et al. (2002)

Table 10.2 CWR abiotic traits transfer to crops- some notable examples

Crop	Traits	Reference
Cicer ssp.	Tolerance to drought and cold	Hajjar and Hodgkin (2007)
	Tolerance to drought and temperature	Eglinton et al. (2001)
	Tolerance to environmental stress	Hajjar and Hodgkin (2007)
	Tolerance to drought and nitrogen stress	Chen et al. (2008)
Lycopersicon ssp.	Tolerance to drought	Rick and and Chetelat (1995)
Manihot ssp.	Tolerance to drought	Jennings (1995)
Musa ssp.	Tolerance to drought	INIBAP/IPGRI (2006)
Oryza ssp.	Tolerance to drought	Zhang et al. (2006)
	Tolerance to acidity, and temperature	Ishimaru et al. (2010)
Triticum ssp.	Tolerance to drought and environmental stress	Nevo (2006)
Zea ssp.	Tolerance to flooding	Mano and Omori (2013)

to maintaining productivity, in response to climate change, because, it introduces novel variation into landrace populations on which selection can act (Sagnard et al. 2011). As a prerequisite to assess the possibilities of gene flow and for proposing natural population conservation strategies, adequate knowledge of the geographical distribution of wild relatives and landrace diversity, in particular in the areas of domestication, is essential. The above table (Table 10.1) gives some important examples of gene transfer between cultivated crops and their wild crop relatives.

The Table 10.2 catalogues examples of the transfer of CWR abiotic traits to crops.

References

Andersson MS, de Vicente MC (2010) Gene flow between crops and their wild relatives. The Johns Hopkins University Press, Baltimore

Castifieiras L, Guzman FA, Duque MC, Shagarodsky T, Cristobal R, de Vicente MC (2007) AFLP's and morphological diversity of *Phaseolus lunatus* L. in Cuban home gardens: approaches to recovering the lost ex situ collection. Biodivers Conserv 16:2847–2865

Chen G, Li C, Shi Y, Nevo E (2008) Wild barley, *Hordeum spontaneum*, a genetic resource for crop improvement in cold and arid regions. Sci Cold Arid Region 1:115–124

Coulibaly S, Pasquet RS, Papa R, Gepts P (2002) AFLP analysis of the phenetic organization and genetic diversity of *Vigna unguiculata* L. Walp reveals extensive gene flow between wild and domesticated types. Theor Appl Genet 104:358–366

Doebley JF, Gaut BS, Smith BD (2006) The molecular genetics of crop domestication. Cell 127:1309–1321

Eglinton JK, Evans DE, Brown AHD (2001) The use of wild barley (*Hordeum vulgare* spp. *spontaneum*) in breeding for quality and adaptation. In: Proceedings of the 10th barley technical symposium Canberra. ACT, Australia, pp 16–20

Eyzaguirre P, Linares O (eds) (2004) Home gardens and Agrobiodiversity. Smithsonian Books, Washington, pp 1–28

Feuillet C, Langridge P, Waugh R (2008) Cereal breeding takes a walk on the wild side. Trends Genet 24:24–32

Galluzzi G, Eyzauguirre P, Negri V (2010) Home gardens: neglected hotspots of agro-biodiversity and cultural diversity. Biodivers Conserv 19:3635–3654

Hajjar R, Hodgkin T (2007) The use of wild relatives in crop improvement: a survey of developments over the last 20 years. Euphytica 156:1–13

Hughes CE, Govindarajulu R, Robertson A, Filer DL, Harris SA, Bailey CD (2007) Serendipitious backyard hybridization and the origin of crops. Proc Natl Acad Sci USA 104:14389–14394

INIBAP/IPGRI (2006) Global conservation strategy for musa (Banana and Plantain). INIBAP/IPGRI, Montpellier/Rome, Italy

Ishimaru T, Hirabayashi H, Ida M (2010) A genetic resource for early-morning flowering trait of wild rice Oryza officinalis to mitigate high temperature-induced spikelet sterility at anthesis. Ann Bot 106:515–520

Jennings DL (1995) Cassava, Manihot esculenta (Euphorbiaceae). In: Smart J, Simmonds JW (eds) Evolution of crop plants. Longman Group, Harlow, Essex, UK, pp 128–132

Mano Y, Omori F (2013) Relationship between constitutive root aerenchyma formation and flooding tolerance in *Zea nicaraguensis*. Plant Soil 370:447–460

Mercer KL, Perales HR (2010) Evolutionary response of landraces to climate change in centres of diversity. Evol Appl 3:480–493

Nevo E (2006) Genetic evolution of wild cereal diversity and prospects for crop improvement. Plant Genet Resour 41:36–46

Nunney L, Campbell KA (1993) Assessing minimum viable population size: demography meets population genetics. Trends Ecol Evol 8:24–239

Rick CM, Chetelat R (1995) Utilization of related wild species for tomato improvement. In: First international symposium on Solanacea for fresh market. Acta Horticulturae, vol 412, pp 21–38

Sagnard F, Deu M, Dembele D (2011) Genetic diversity, structure, gene flow and evolutionary relationships within the Sorghum bicolor wild-weedy-crop complex in a western African region. Theor Appl Genet 123:1231–1246

Zhang X, Zhou S, Fu Y, Su Z, Wang X, Sun C (2006) Identification of a drought tolerant introgression line derived from Dongxiang common wild rice (*Oryza rufipogon* Griff.). Plant Mol Biol 62:247–259

Chapter 11
In Situ Conservation Research in CWR

The preferred approach in CWR conservation is in situ research. The advantage in this approach is that the target species are continuously exposed to a changing natural environment which allows new diversity to be generated (Hunter and Heywood 2011). The principal objective and long-term goal of in situ conservation of species is to protect, manage, and monitor the populations in their natural habitats so that the natural evolutionary processes can be maintained, and allowing in this way, the creation of new variation in the genes which enable the species to adapt to gradual changes in environmental conditions (Heywood and Dalloo 2005). This approach has gained increasing attention globally as evidenced by their inclusion in the many national reports compiled for the SOW2 (FAO 2010). Notwithstanding the increase in isolated activities targeting CWR conservation, the formal recognition and/or the adoption of appropriate management regimes to protect CWR in situ are mostly lacking. Unfortunately, no quantitative data are provided by countries on the changing status of CWR, but, several reports indicated that specific measures had been taken to promote their conservation. During the last decade, the number and coverage of protected regions have increased approximately by one—third, (MDG Report 2010), yet, only limited efforts have been made to target CWR, whose conservation remains unplanned and largely an indirect effect of protecting flagship species of threatened habitats. CWRs have only been passively conserved within existing protected areas, but, their conservation is not guaranteed and it is almost certain that individual populations may possibly decline as time passes or even be lost entirely (Maxted and Kell 2009). There are, of course, a limited number of protected areas which have been established specifically for CWR as genetic reserves. The wild emmer wheat (*Triticum turgidum* L.var.*dicoccoides*) in the Ammiad reserve in the eastern Galilee, in Israel, is one such example (Anikster et al 1997), and a perennial wild relative of maize (*Zea diploperennis* H.H.I) in the Man and Biosphere Sierra de Manantlan Biospehre Reserve endemic to southwest Mexico (UNESCO 2007). Various cereal, forage, and fruit trees in CWR reserves were established in Lebanon, Syria, Palestinian Territories, and Jordon (Amri et al. 2008 a), wild wheat in the Erebuni reserve in Armenia, wild *Coffea* species in the Mascarene Islands (Dulloo et al. 1999).

© Springer Nature Switzerland AG 2019
K. P. Nair, *Combating Global Warming*, Springer Climate,
https://doi.org/10.1007/978-3-030-23037-1_11

Heywood and Dulloo (2005) reported that conservation of CWR outside protected areas is more problematic. Many CWR of major crops are mostly found in disturbed, preclimax plant communities, and therefore are located outside protected areas. And these sites are not managed well for biodiversity conservation, and hence, the occurrence of CWR population is only incidental, which makes them particularly vulnerable to adverse management changes. The Dryland Agrobiodiversity Project in West Asia found that many intensively cultivated areas contain significant CWR diversity at their margins in field edges, habitat patches or roadsides (Al-Atawneh et al. 2008). In the base of the Beqaa Valley, Lebanon, which is industrially cultivated, there are globally significant populations of rare CWR found along roadsides, while in the Hebron area of Palestine and Jabal Al-Druze in Syria, very rare wheat, lentil, barley, pea, and bean CWR are common in modern apple orchards. There is an urgent need for prompt in situ conservation of Teosintes in maize putative center of origin, in the Balsas, Guerrero in Mexico. In another investigation, Ureta et al. (2012) pointed, in particular, to *Tripsacum intermedium* (De Wet and J.R.Harlan) and *Zea perennis* (Hitchc.) (Reeves and Mangelsd) as the most vulnerable wild relatives of maize and listed a few landraces which may require special attention. It is important to establish some level of protection for these sites, with consistent management, which can be reached through an agreement with the owner of the site where these endangered CWRs are located. A well—documented example of these kinds of local management agreements are those used in the establishment of microreserves in the Valencia region of Spain (Serra et al. 2004). The "100 fields for biodiversity" project in Germany aims to manage the identified fields in a manner conducive for the growth and establishment of rare CWRs (www.schutzaecker.de).

It is unfortunate that in a country like India, where there are CWR hotspots for rice in states like Odisha, in eastern India, there has been a silent siphoning of some of the rare varieties out of the country by unscrupulous scientists, for pecuniary and personal professional benefits, thanks to the green revolution. The Indian Council of Agricultural Research, the apex body which governs the country's outreach research programs in various crops is still to wake up to a new and urgent reality of CWR conservation. What happened in India is beautifully illustrated in the article "The Great Gene Robbery" (Claude Alvarez, The Illustrated Weekly of India 1986).

References

Al-Atawneh N, Amri A, Assi R, Maxted N (2008) Management plans for promoting in situ conservation of local agrobiodiversity in the West Asia centre of plant diversity. In: Maxted N, Ford-Lloyd BV, Kell SP, Irinondo J, Dulloo ME, Turuk K (eds) Crop wild relative conservation and use. CABI Publishing, Wallingford, pp 338–361

Alvarez C (1986) The great gene robbery. The Illustrated Weekly of India. March 23, 1986

Amri A, Monzer M, Al-Oqla A, Atawneh N, Shehadeh A, Konopka J (2008) Status and treats to natural habitats and crop wild relatives in selected areas in West Asia region. In: Proceedings of the international conference on promoting community driven in situ conservation of dryland Agrobiodiversity ICARDA, Aleppo, Syria

Anikster Y, Feldman M, Horovitz A (1997) The ammiad experiment. In: Maxted N, Ford-Lloyd BV, Hawkes JG (eds) Plant genetic conservation: the in situ approach. Chapman and Hall, London, pp 239–253

Dulloo ME, Guarino L, Engelmann F (1999) Complementary conservation strategies for the genus Coffea with special reference to the Mascarene Islands. Genet Resour Crop Evol 45:565–579

FAO (2010) The second report on the state of the world's plant genetic resources for food and agriculture. FAO, Rome, Italy, p 370

Heywood VH, Dulloo ME (2005) In situ conservation of wild plant species: a critical global review of best practices. IPGRI Technical Bulletin II. IPGRI, Rome, p 174

Hunter D, Heywood V (2011) Crop wild relatives: a manual of in situ conservation, 1st edn. Earthscan, London, UK

Maxted N, Kell SP (2009) Establishment of a global network for the in situ conservation of crop wild relatives: status and needs. Italy, FAO Commission on Genetic Resources for Food and Agriculture, Rome, p 266

MDG Report (2010) United Nations (2010) Millennium Development Goals Report 2010 USA. United Nations. ISBN: 978-92-1-101218-7

Serra L, Perez-Rovira P, Deltoro VI, Fabregat C, Laguna E, Perez-Botella J (2004) Distribution, status and conservation of rare relict plant species in the Valencian community. Bocconea 16:857–863

UNESCO (2007) UNESCO-MAB Biosphere Reserves Directory Biosphere Reserve Information. The MAB Programme United Nations Educational, Scientific and Cultural Organization, Mexico, Sierra de Manantlan. http://wwwunesco.org/mabdb/br/brdir/directory/bioresasp?

Ureta C, Martinez-Meyer E, Perales HR, Alvarez-Buylla ER (2012) Projecting the effects of climate change on the distribution of maize races, and their wild relatives in Mexico Global Change. Biology 18:1073–1082

Chapter 12
Ex Situ Conservation Research in CWR

Both in situ and ex situ conservation research must be looked at as complementary to each other (Dulloo 2011). The CWRs are poorly represented gene banks and there are many examples to show this. There are 7.4 million accessions known to be conserved ex situ. Of these, only 2% to 18% of ex situ collections are CWR (Maxted et al. 2012). In the USA, over 96,000 gene bank accessions of 2800 taxa listed in the Inventory are recorded in Germplasm Resources Information Network (GRIN), but, a large proportion of this material is cultivated germplasm conspecific with wild taxa, such as, American cotton (*Gossypium hirsutum* L.) and chili pepper (*Capsicum annum* L.). As discussed earlier, many CWRs are threatened by climate change and other anthropogenic actions, such as, habitat degradation, fragmentation, deforestation, etc. And quite often, it is impossible to reverse these threats, and the collection and conservation in ex situ collections is the only method to safeguard them. It is not always cost-effective to conserve all the populations of CWR in the wild. The national CWR strategies and key national CWR protected areas should include a safety backup to ensure the conservation of the germplasm (Maxted and Kell 2009). Such strategies should collectively prioritize collection as an important activity for the present and future, both for wild relatives and landraces. The Global Crop Variety Trust, with support from the Government of Norway, launched a project in partnership with the Millennium Seed Bank of the Royal Botanic Gardens, Kew, and a number of country partners to identify and collect CWR of 29 crops of major importance to food security which are missing from existing collections, for their safe conservation banks (http://www.cwrdiversity.org/about-us/). The collected material will then be evaluated for important traits and prepared for use in adapting crops for new climates through pre breeding programs. In Israel, Hubner et al. (2012) established an ex situ collection of wild barley which contains 1020 wild barley populations from 51 sites that capture barley diversity at multiple levels in Israel from microsite to country-wide areas. In China, more than 5000 wild rice samples, representing four taxa of *Oryza* and seven species in the related genera of the *Oryzeae* collected during 1978–82 in different provinces of China, have been deposited in the Chinese National Genebank in Beijing, as well as gene banks and seed storage facilities of provincial and local agricultural agencies (Lu 2012).

© Springer Nature Switzerland AG 2019
K. P. Nair, *Combating Global Warming*, Springer Climate,
https://doi.org/10.1007/978-3-030-23037-1_12

One of the major concerns of ex situ conservation research pertains to the seed quality. Orthodox seeds are the most frequently used materials for ex situ conservation, as they have the capability to remain viable for long, when they are dried and stored at low temperatures. However, current research shows that variability exists in seed longevity/viability for different species being conserved under similar conditions. When one wants to collect CWR, which comprise a wide diversity of species, the above issue would be of great relevance for most gene banks, which conserve single crops. Seeds of plants which cannot be conserved because of their desiccating and cold sensitive nature, or which are clonally propagated, are traditionally conserved as live plants in field gene banks (Dulloo 2011). However, there are logistical problems with field gene banks, such as: large space requirement, financial outlays, vulnerability to pests and diseases, natural disasters, extreme weather, fire, theft, vandalism, and the risk because of changes in administrative policy of the concerned government where the gene bank is located. Research to tackle some of these problems has shown that these difficult to store seeds and clonally propagated species can be stored employing in vitro cryopreservation methods. The latter technique is discussed in the book on turmeric and ginger (Nair 2013). The in vitro slow-growth method involves maintenance of explants in a sterile, pathogen-free environment and has been applied to over 1000 different species, although, individual maintenance protocols need to be developed for the majority of species (Thormann et al. 2006). Significant progress has been made in cryopreservation research over the past twenty years, and much of that research has been focused on understanding the desiccation sensitivity of recalcitrant seeds and on the underlying mechanism of desiccation tolerance (Engelmann and Panis 2009). Ex situ conservation of CWR can benefit from the progress which is being made in the development of those conservation techniques. Further, additional regeneration challenges are also presented by ex situ conservation of CWR (Engels and Rao 1998). Many wild species, including those with self-pollinating (autogamous) domesticated relatives, are cross-pollinating, as in wheat (*Aegilops speltoides* Tausch) (Zaharieva and Monneveux 2006), barley (*Hordeum bulbosum* L.), and oat (*Avena macrostachya* Balensa ex.Coss. et Durieu). They require adequate isolation from related species during regeneration to avoid gene flow between species. Also, certain CWR, such as wild oat (*Avena fatua* L.) has the potential to become noxious weeds and, therefore, require special containment measures during regeneration. Additionally, the question of effective conservation of genetic diversity of CWR is the minimum number of wild populations of CWR from which seeds should be collected and the number of seeds which should be collected. Ex situ is also recommended as a means of distribution of CWR to gene banks and breeders in other countries who otherwise may not have access to specific "in situ" populations. It also provides a reference of within population diversity at a given time period.

References

Dulloo ME (2011) Complementary conservation actions. In: Hunter, D, Heywood DH (Eds), Crop wild relatives: a manual of in situ conservation. Earthscan, London, Washington DC, pp 275–294. (Chapter 12)

Engelmann F, Panis B (2009) Strategy of biodiversity international cryopreservation research—a discussion paper. Biodiversity International, Rome, Italy, 8 pp (unpublished)

Engels JMM, Rao RR (1998) Regeneration of seed crops and their wild relatives: Proceedings of a consultation Meeting, 4–7, December 1995, ICRISAT, Patancheru, Hyderabad, Andhra Pradesh, India. IPGRI, Rome, Italy

Hubner S, Gunther T, Flavell A (2012) Islands and streams: clusters and gene flow in wild barley populations from the Levant. Mol Ecol 21:1115–1129

Lu BR (2012) The challenge of in situ conservation of crop wild relatives in the biotechnology Era: a case study of wild rice species. In: Maxted N, Dulloo ME, Ford-Lloyd BV, Frese L, Iriondo J, Pinheiro de Carvalho MAA (Eds), Agrobiodiversity conservation: securing the diversity of crop wild relatives and landraces. CAB International Publishing, Wallingford, UK, pp 211–217

Maxted N, Kell SP (2009) Establishment of a global network for the in situ conservation of crop wild relatives: status and needs. FAO Commission on Genetic Resources for Food and Agriculture, Rome, Italy, p 266

Maxted N, Dulloo ME, Ford-Lloyd BV, Frese L, Iriondo J, Pinheiro de Caravalho MAA (2012) Agrobiodiversity conservation: securing the diversity of crop wild relatives and landraces. CABI

Nair KPP (2013) The buffer power concept and its relevance in African and Asian soils. Adv Agron 121:416–516

Thormann I, Dulloo ME, Engels JMM (2006) Techniques of ex situ plant conservation. In: Plant conservation genetics. Centre for Plant Conservation Genetics, Southern Cross University. The Haworth Press Inc, Lismore, Australia, pp 7–36

Zaharieva M, Monneveux P (2006) Spontaneous hybridization between Bread wheat (Triticum aestivum L.) and its wild relatives in Europe. Crop Sci 46:512–527

Chapter 13
Utilizing CWRs in Major Food Crops to Combat Global Warming

It is almost a century back that attempts to use genes from crop wild relatives (CWRs) to improve crop production were initiated. However, it is during the past six to seven decades that the rate of release of cultivars containing genes of wild relatives was exploited in crop breeding programs to impart resistance and/or tolerance to biotic factors such as diseases and pests. However, the utilization of CWRs to combat abiotic stresses such as the environmental hazards of global warming is a relatively new endeavor, or imparting resistance and/or tolerance to frost, drought, acidity, salinity, etc. Genetic compatibility between cultivated material and wild species may, in fact, make the utilization of the latter straightforward, but, equally, in some cases, genetic incompatibility may make the task more complicated. The following discussion will focus on the attempts where the use of wild relatives of some major food crops in combating biotic and abiotic stress of cultivated crops, especially against the background of the looming shadow of global climate change, have been attempted. The sheer size of the monetary gain in utilizing CWRs in crop production is staggering—the yield and quality contributions add to in excess of US $ 340 million to North American economy (Presscott-Allen and Presscott-Allen 1986). This discussion covers the following crops:

13.1 Wheat

It was in the Fertile Crescent (in the Israeli region) that wheat happens to be one of the earliest crops to be domesticated, where *Triticum* species and their close relatives exhibit enormous diversity. The *Triticum* genus includes both diploid and polyploid species. *Tritcum turgidum* ssp. durum belongs to the tetraploid species, which arose from a cross between the diploid *Triticum urattu* and a diploid *Aegilops* species belonging to the Sitopsis section. Tetraploid wheat, when later crossed with wild goat grass *Aegilops tauschii*, gave rise to the hexaploid species among which is the common wheat *Triticum aestivum.* Wheat forms the staple food for the world population, hence, to step up it's production, it becomes imperative that its genetic

© Springer Nature Switzerland AG 2019
K. P. Nair, *Combating Global Warming*, Springer Climate,
https://doi.org/10.1007/978-3-030-23037-1_13

base be broadened through the introduction of genes from nonconventional sources such as crop wild relatives (CWRs), its progenitors, as well as obsolete forms of the crop no longer cultivated in large quantities. This becomes all the more important against the increasing environmental hazards which stem from global warming. The recent cyclone "*Ockhi*" in Kerala State, in 2017, that left a trail of misery in the state's coast, in India, and in 2016 the tsunami in Tamil Nadu, also in India, both untimely, which rendered thousands of people dead, in addition to causing huge environmental hazards, are grim warnings that attempts to boost food production can no longer be deferred. Urgent and foolproof measures are required.

The need to tap wild relatives or progenitors of wheat to boost wheat yield must have been in the mind of early wheat breeders in North America and Europe. The first cross between *Triticum aestivum* and *Triticum villosum* (synonymous with *Dasypyrum villosum*) in the spring of 1908 and the first variety under the name "Cantore" was released in 1919 (Strampelli 1932). However, many of those early attempts were frustrated because of crossability barriers, differing ploidy levels, genomic non-homology, and hybrid sterility. Even in cases where crosses were made, segregation in subsequent generations rendered the results without promise (Kimber 1993). In the 1940s, additions of wild species single chromosomes to wheat were made, and it was evident that the negative effects of entire genome incorporation could be mitigated to some extent. However, the cytological stability of disomic alien additions in commercial wheat cultivars was limited, and the alien material was rapidly lost in subsequent generations. This problem was eliminated when the alien chromosome was substituted for its wheat homoeologous. It was concluded that the smaller the segment of alien chromosome introduced, the greater the chances of successfully producing a commercial variety. The first attempt to make a small alien insertion into cultivated wheat was carried out by Sears (1956) when this author successfully translocated part of a chromosome of *Aegilops umbellulata* in the wheat chromosome 6B.

The use of wild relatives to contribute resistance/tolerance to abiotic stresses and in increasing crop harvests and quality has been a problematic endeavor. Many wild wheat relatives have been described as potentially useful in contributing genetic resistance/tolerance to abiotic stresses, although only a handful of examples of progenies from crosses between cultivated and wild material have reached the stage of cultivar release (Shannon 1997). In general, these results were not unexpected, as the poor agronomic performance of wild relatives is well known and hence, the difficulties in recovering high yield potential. The attempt to create synthetic hexaploid wheat by crossing *durum* wheat with the wild parent *Ae.tauschii*, followed by chromosome doubling (Mujeeb-Kazi et al. 1996) is a noteworthy example. From a series of backcrosses of this material with elite common wheat varieties, it was possible to release the variety "Chuanmai 42", which produces about 20–35% more yield than the original crop variety, in addition to possessing beneficial traits such as tolerance to water-logging (Villareal et al. 2001), spot blotch (*Cochliobolus sativus*), and "karnal bunt" which was successfully done in 2003. It has been also noted that there is an increase in the protein content in cultivars derived from crosses with *T.dicoccoides* (Hoisington et al. 1999), although the definition of the genome contributing the genes involved was not attempted.

13.1.1   Oryza Sativa

Rice (*Oryza sativa*) is an equally important staple food for humans as wheat. About twenty wild rice species could be considered as genetic resources in the rice gene pool (Chang 1970). The International Rice Research Institute (IRRI) has used wild relatives of rice to enhance grain quality and also to incorporate high levels of resistance to major pests and diseases. This work includes the transfer of a dominant gene for resistance to the grassy stunt virus from the wild relative *Oryza nivara* (Beachell et al. 1972). Review results on the use of wild rice, specifically to combat global warming, shows that there is but scanty useful results, though, there are other numerous examples where resistance to pests and diseases have been successful through wide rice introgression. The only outstanding example is of *Oryza nivara*, a wild rice species with an AA genome, which is the sole donor of a major dominant gene to confer resistance to grassy stunt virus biotype-1 which also possesses resistance to sheath blight (Barclay 2004). This wild relative of modern cultivated Asian rice also possesses other important agronomic traits such as resistance to blast, stem rot, drought avoidance, and Cytoplasmic male sterility (Brar and Khush 2003).

13.1.2   Zea Mays

Maize (*Zea mays*), or corn, contributes more than 20% of human food worldwide. It was the native Americans who domesticated nine of the most important food crops in the world, of which maize is one. Maize has been linked to Teosinte, a Mexican grass (Beadle 1939), and its domestication dates back to about 9000 years ago. Considered to be phenotypically the most distinctive, as well as the most threatened teosinte, is *Zea nicaraguensis*, which thrives in flooded conditions along 200 m of a coastal estuarine river in northwest Nicaragua. Despite the wide diversity of maize, there hardly are any worthwhile examples of the utilization of its wild relatives in corn breeding, especially to combat global warming. But, it is a sure bet that both maize and teosinte, its cousin, will remain in global farming, the latter providing much fodder for milch cows and the former the food for humans. Both are indistinguishable, one from the other, until about the tasseling (male flower blooming) time. Thanks to recent advances in maize genomics, it is now possible to undertake candidate-gene-based association mapping which could be a promising method for investigating the inheritance of complex traits in teosinte (Prasanna 2012).

13.1.3 Solanum Tuberosum

Potato (*Solanum tuberosum*) has an interesting history. When Charles Darwin reached Guayateca island on the Chilean archipelago of Los Chonos, he noted a luxurious growth of wild potatoes. The plants were tall, the tubers small, and oval shaped resembling the English potato (Darwin 1839). In 1969, almost a century and a half later, the Peruvian plant explorer Carlos Ochoa entered a cave on the same island, found the same potato described by Darwin. In honor of Ochoa's discovery, the species was named *Solanum tuberosum*. He postulated that this potato had at some time been cultivated and then grew wild, because it features the same chromosomes and a similar morphology to *Solanum tuberosum* (the cultivated potato). There is no published report which suggests the use of wild potatoes to combat global warming. There are a number of reports to show that it can be used to confer resistance to certain diseases, especially the notorious virus disease, the potato virus Y (PVY). Hijmans et al. (2003) investigated several species of wild potatoes, assessing the predictivity of taxonomic, geographic, and ecological factors in ascertaining which of these species was suitable as gene donors for cold tolerance. They found that *Solanum acaule, Solanum albicans,* and *Solanum commersonii,* respectively, were most tolerant to frost. The wild potato is an endangered plant, and most species are rare and difficult to find. There are about 93 different species in Peru and 39 in Bolivia.

13.1.4 Cicer Arietinum

Chickpea (*Cicer arietinum*) has been under domestication during more than 7500 years in the Near East. It is one of the most nutritious food crops of the Middle East and much of South Asia. Also found in the Americas, both North and South, it is known as Garbanzo beans. Attempts to increasing Chickpea's yield potential have been aplenty but there is only scanty attempt to use its wild relatives to boost yield potential. One noteworthy attempt has been of Singh and Ocampo (1997) who managed to increase the yields of cultivated chickpea, by utilizing genes from two wild species, namely, *Cicer echninospermum* and *Cicer reticulatum*. The improved lines were tested for cooking quality as well as increased yields and were found to have no difference when compared to the standard cultivated chickpea varieties available in the Middle East. Nor did the lines have any of the undesirable traits of the two wild species. *Cicer microphyllum,* another wild relative of cultivated chickpea, is a high-altitude cold desert-adapted species distributed in the western and trans-Himalayas.

13.1.5 Lens Culinaris

Lentil (*Lens culinaris* subsp. *Culinaris*), the genus *Lens*, belongs to the family Fabaceae, includes the cultivated *Lens culinaris* subsp.*Culinaris*, the wild subspecies *L.culinaris* subsp. *orientalis*, the progenitor *L.culinaris* subsp. *tomentosus* and *L.culinaris* subsp. *odemensis,* along with three other wild species. The taxa within *culinaris* are in the primary gene pool, while *Lens ervoides*, *L.nigricans,* and *L.lamottei* are in the secondary–tertiary gene pools. The taxonomy of *Lens*, however, remains debatable. All *Lens* species are self-pollinating annual diploids (2n = 14). The oldest carbonized remains of lentil are from Franchthi cave in Greece dated back to 11,000 BC and from Tell Mureybit in Syria dated back to 8500–7500 BC. But, as it is not possible to differentiate the wild from the cultivated small-seeded lentil, the state of domestication of these and other carbonized remains in the aceramic farming villages in the seventh millennium BC in the Near East are unknown. Information from several investigations during the past quarter century indicate that genetic variation present in *L.orientalis* and *L.nigricans* can be easily exploited for crop improvement. However, the full extent of genetic variation is still relatively unknown as they are poorly represented in world collections, so also their usefulness, from the wild source, in deriving global warming resistant types.

13.1.6 Utilization of Wild Relatives in the Breeding of Tomato and Other Major Vegetables

Reliance on just three principal cereals like rice, wheat, and maize and a few other carbohydrate-rich staples might be sufficient to attain, what generally is known as "food security" but the term "nutritional security" has another dimension. They form a large and very diverse commodity group and vegetables are an important source of essential vitamins, antioxidants, minerals, fiber, amino acids, and other health-promoting compounds. Vegetables form a large and very diverse commodity group which include a wide range of genera and species. The United Nations Food and Agriculture Organization (FAO, Rome) tracks a group of twenty-seven commodities under "vegetables and melons", which had a global production of more than 1 billion tons in 2010 (FAOSTAT 2013). More than 75% of vegetables are melons produced in Asia (794.3 million tons, amounting to 76.05% of the total), followed by Europe (94.2 million tons, 9.02% of the total), Americas (81.2 million tons, 7.77% of the total), African continent (71.2 million tons, 6.81% of the total), and Oceania (3.6 millions tons, 0.34% of the total).

The Asian Vegetable Research and Development Center (AVRDC), which is the world vegetable center, houses the world's largest sector collection of vegetable germplasm. It has assembled more than 61,000 accessions of vegetables covering 172 genera and 440 species from 155 countries. Most of these accessions are either landraces or wild relatives of the cultivated forms and represent a unique and invaluable resource for Plant Breeders worldwide. Vegetable germplasm collections of selected major crops held worldwide have recently been described (Ebert 2013).

Crop Wild Relatives are a vast source for vegetable improvement focusing on biotic and abiotic stresses caused by climate change. While increased ambient temperature might lead to higher vegetable production in some countries of the world, in others the effect might be just the reverse, affecting both productivity and quality of the produce (de la Pena et al. 2011). Additionally, climate change will lead to the development of more predators, like new fungal and bacterial pathogens. Also, elevated temperatures induce excessive water evaporations from the soil surface and speed up the plant development which leads to lessened productivity (Battisti and Naylor 2009). Drought can lead the plant to display different mechanisms to withstand the stress, including shortening of the life cycle of the plant, improving water uptake and reducing transpiration, and tissue tolerance to dehydration (Chaves and Oliveira 2004).

Exploitation of genetic diversity to develop stress-tolerant crops is of strategic importance to combat the negative impact of climate change on crop production (de la Pena et al. 2011). Since ancient times CWRs have served as the basis for crop domestication and improvement. Today, CWRs which are threatened in the wild, which are only partially conserved, in gene banks, have been rediscovered as an essential resource for crop improvement programs to combat both biotic and abiotic stresses. Against this background, tomato emerges as a model crop in research and breeding for which an enormous amount of biotic and abiotic stress tolerance has already been investigated in the entire pool of its CWRs. This might have to do with the domestication bottlenecks which reduced the genetic diversity of cultivated tomatoes to less than 5% of that of its CWRs (Miller and Tanksley 1990). Much less has been investigated with regard to the potential of the CWRs in other vegetable crops.

Economic and Nutritional Importance of Tomato: Domesticated tomato, *Solanum lycopersicum*, is produced and consumed worldwide and it is the world's most important vegetable commercially. The health benefits, arising out of the phytonutrients it contains, have generated global scientific and commercial interest. Cultivated tomato is a model crop for genetic, developmental, and physiological studies. Given its relatively small genome size, diploid genetics, short reproductive cycle, and a great diversity of genetic resources, tomato genome has been selected as one of the model genomes for the *Solanaceae* family, and international genome sequencing efforts led to the publication of the tomato genome in 2012 (The Tomato Genome Consortium 2012; Mueller 2013).

In terms of net production value, cultivated tomato (*S.lycopersicum* L.,) is the most important vegetable grown worldwide, ranked eighth among all agricultural commodities in 2010 (FAOSTAT 2013). In 2010, global production reached 152 million tons with a net production value of US $55.6 million. The top five producers were China (46.9 million tons—mt), US (12.9 mt), India (12.4 mt), Turkey (10.1 mt), and Egypt (8.5 mt). During 1991–2010, the tomato production area expanded from

2.86 million hectares (mha) to 4.53 mha −58.4% increase, while the yield increased from 266.146 to 335.487 kg/ha (26.1% increase, and total quantity produced from 76.09 to 152.06 mt—more than a 100% increase FAOSTAT 2013).

Compared to other leafy vegetables, tomato is not a nutrient-dense food source (Keatinge et al. 2011). However, because of the relatively large quantities consumed, tomatoes make a substantial nutritional contribution to the human diet. In the US, the world's second-largest producer, next to China, tomato is the fourth most popular food crop along with potato, lettuce, and onions (USDA-ERS 2013). Tomato fruits contain considerable quantities of β carotene, a provitamin A carotenoid, and ascorbic acid. Apart from their value as a provitamin and vitamin, respectively, beta carotene and ascorbic acid also function as antioxidants (Hanson et al. 2004).

The principal carotenoid in tomato is lycopene, which does not have retinoid activity but serves as a powerful antioxidant. Lycopene-rich diets have been associated with lower risks of certain cancers, cardiac diseases, and old-age related health problems. Consumption of tomato and tomato products has been shown to protect the DNA from oxidative damage which can lead to cancer (Ellinger et al. 2006). As it has proven difficult to establish a direct association between lycopene and cancer incidence, it has been suggested that a mix of several phytonutrients found in tomato have additive or synergistic effects in promoting good health and reducing disease risks (Ellinger et al. 2006). The positive effect of tomato consumption on human health extends beyond the positive effect as an antioxidant and includes antithrombotic and anti-inflammatory functions (Burton-Freeman 2011). Given its popularity and worldwide availability in multiple forms—fresh, canned, sauce, and dried—encouraging greater tomato consumption in fresh and processed forms might be an effective way of increasing overall vegetable intake, ultimately leading to better human health.

Origin, domestication, and dissemination of tomato:

Tomato originated in the South American Andes, ranging from northern Chile in the south, through Bolivia, Peru, to Ecuador and Colombia in the north (Bai and Lindhout 2007). Two wild species, *Solanum galapagense* and *Solanum cheesmaniae*, are endemic in the Galapagos (Peralta et al. 2008). Wild tomatoes are found in a wide range of habitats in western South America, from sea level to elevations above 3600 m.

Six wild tomato species, *Solanum chilense, Solanum habrochaites, Solanum pennellii, Solanum peruvianum, Solanum arcanum, and Solanum pimpinellifolium*—are found in the arid Pacific coastal lowlands and adjacent hills (Perlata et al. 2008). Other species have their habitats in valleys of rivers leading into the Pacific or in the uplands of the Andes. The valley of Rio Maranon is the main habitat of *S.arcanum*. *S.huaylasense* is endemic in the valley of Rio Santa, while *Solanum corneliomuelleri* is in valleys from central to southern Peru. *Solanum chmielewskii* is in the upper Apurimac valley of Peru and the Sorata valley of Bolivia, and *Solanum neorickii* in the dry valleys extending from Ecuador to southern Peru. Four species—*S.chilense, S.corneliomuelleri, S.pennellii, and S.habrochaites*—are also found in the high altitudes of the Andes (Peralta et al. 2008). In the Galapagos Islands, *S.galapagense* commonly grows at lower elevations, while *S.cheesmaniae* thrives from sea level to the rocky slopes of volcanoes.

Initially, Peru had been proposed as the center of domestication of cultivated tomatoes (De Candolle 1886). This would coincide with its center of origin and genetic diversity but Peru lacks depictions of this crop on textile or pottery artifacts during the pre-Colombian era (Rick 1995). Linguistic evidence pointed to Mexico and Central America as the center of domestication of tomato as the word tomato has its origin in the Aztec word *xitomatl* (Cox 2000) and tribes in Central America called the crop tomati (Gould 1983). The ancient Peruvian tribes do not mention a tomato-like fruit at all, while Aztec documents in Central America contain records of meals with peppers, salt, and tomato (Cox 2000). Genetic evidence also pointed to Mexico as the center of domestication as modern cultivars appeared to be more closely related to a cherry tomato-like cultivar grown widely in Mexico and throughout Central America at the time of the discovery by the Spanish than to any wild species grown in Peru (Rick 1995). The cherry tomato (*S.lycopersicum* L., var. *cerasiforme* (Alef.) Fosberg) is likely the direct ancestor of cultivated tomato (Tansley 2004), and this botanical variety is still found in a semi-wild state in Central America. The popularity and worldwide production and consumption of tomato is at least partially due to the Spanish explorers who discovered the crop in Mexico and Central America and introduced it to Spain in the early sixteenth century under the name *pome dei Moro* (Moor's apple). From Spain, the crop spread to Italy and France, where the fruit was called *pomme d'amour* (love apple), possibly a corruption of the early Spanish name *pome dei Moro* (Cox 2000). The Spanish distributed tomato throughout their colonies, including the Philippines, from where it reached the other parts of Asia. It was only in the nineteenth century that tomatoes became widely accepted in the US, and this culminated into a brief "tomato fever", as medicinal powers were attributed to the fruit and tomato extract was added to almost every pill as a panacea (Cox 2000). Although of short duration, this tomato mania boosted the popularity of the crop enormously.

Taxonomy, phylogenetic relationships, ploidy level and mating system:

Tomato, pepper, and potato are important and widely cultivated members of the *Solanaceae* family, which comprises 95 accepted genera (USDA-ARS 2013), and more than 3000 species. All three crops originated in the New World—the Central and South America (Knapp 2002). Despite the enormous number and global distribution of the genera and species included in the family, cytogenetically, it is a very conservative family as most taxa are characterized by a basic chromosome number of $n = 12$ (Chiarini et al. 2010). There has been much debate on the generic status of tomato since the sixteenth century when the crop was introduced by the Spaniards into Europe. Botanists noted tomato's close relationship with the genus *Solanum* and referred to this species as *Solanum pomiferum* (Luckwill 1943). The first taxonomist to assign the generic name Lycopersicon to tomato was Joseph Pitton de Tournefort (1694 cited by Peralta et al. 2006). In 1753, Carolus Linnaeus grouped tomato under the genus *Solanum*, but a year later Philip Miller followed the nomenclature of

Table 13.1 Taxonomic classification of tomato and its wild relatives (section *Lycopersicon*) with the two closely related sections *Lycopersicoides* and *Jugalandifolia*

Section	Group	Species	Breeding system
Lycopersicon	*Lycopersicon*	S.*lycopersicum*	SC[a], autogamous, facultative allogamous
		S.*pimpinellifolium*	SC[a], autogamous, facultative allogamous
		S.*cheesmaniae*	SC, exclusively autogamous
		S.*galapagense*	SC, exclusively autogamous
	Neolycopersicon	S.*pennellii*	Usually SI[b], some SC
	Eriopersicon	S.*habrochaites*	Typically SI, some SC
		S.*huaylasense*	Typically Si, allogamous
		S.*corneliomuelleri*	Typically SI, allogamous
		S. *peruvianum*	Typically SI, allogamous
		S.*chilense*	SI, allogamous
	Arcanum	S.*arcanum*	Typically SI, allogamous rarely SC
		S.*chmielewskii*	SC, facultative allogamous
		S.*neorickii*	SC, highly autogamous
Lycopersicoides		S.*lycopersicoides*	SI, allogamous
		S.*sitiens*	SI, allogamous
Jugaldifolia		S.*jugalandifolium*	SI, allogamous
		S.*ochranthum*	SI, allogamous

Source Adapted from Robertson and Labate (2007), Peralta et al. (2008), Caicedo and Peralta (2013)
[a]SC = Self compatible
[b]SI = Self incompatible

Tournefort and described tomatoes formally under the genus *Lycopersicon* (Peralta et al. 2006). In a posthumous edition of the book The Gardener's and Botanist's Dictionary (Miller 1807 cited by Peralta et al. 2006), the book editor decided to follow Linnaeus nomenclature for tomato and merged the genus *Lycopersicon* with *Solanum*, describing the tomato species as *Solanum lycopersicum*. Today, tomatoes are formally classified under the genus *Solanum* sect. *Lycopersicon*. This classification is based on evidence derived from phylogenetic studies using DNA sequences and more in-depth studies of plant morphology and distribution of the species (Peralta et al. 2006).

The relatively small section *Lycoepersicon* in the genus *Solanum* comprises one domesticated species *Solanum lycopersicum* and 12 Crop Wild Relatives, which are detailed in the following table (Table 13.1, Peralta et al. 2008). Two species in *Solanum* section *Juglandifolia* are sisters to section *Lycopersicon* which are found in Colombia, Ecuador, and Peru. Another two species in *Solanum* section *Lycopersicoides* are sisters to sections *Lycopersicon* and *Juglandifolia*. The latter have their habitats in southern Peru and northern Chile (Peralta et al. 2008).

Most species within the *Lycopersicon* group can reciprocally hybridize with cultivated tomato, with the exception of *Solanum habrochaites* (Robertson and Labate 2007). *Solanum habrochaites* can act as a pollen parent in crosses with cultivated tomato but the reciprocal cross does not set fruit. Within the *Eriopersicon* group, compatibility with cultivated tomato is rather limited. *Solanum chilense* can act as pollen parent for *Solanum lycopersicum* but stable seeds are rare (Robertson and Labate 2007). The reciprocal cross is not possible as *Solanum chilense* does not accept pollen from the cultigen. *Solanum peruvianum* presents severe crossing barriers in hybridization attempts within cultivated tomato.

It has been estimated that the genomes of the tomato cultigens contain less than 5% of the genetic diversity of their wild relatives (Miller and Tanksley 1990). Apparently, the domestication and transmigration process of tomato from the Andes to central America and from there to Europe caused a major genetic drift in this inbreeding cultigen. Despite this narrow genetic base, cultivated tomato is extremely rich in shapes, colors, and sizes, in contrast to the wild forms which bear only tiny fruit. It is likely that mutations associated with larger fruit were selected and accumulated during tomato domestication (Bai and Lindhout 2007). Only in the twentieth century did the genetic potential of wild tomato relatives become apparent in initial crosses made with cultivated tomato by plant geneticist and botanist Charles Rick who established and guided the C. M. Rick Tomato Genetic Resource Center at the University of California, Davis campus, in USA. Interspecific crosses are now widely used to tap into the gene pool of wild tomato relatives when breeding for resistance to biotic and tolerance to abiotic stresses.

Achievements with classical tomato breeding using Crop Wild Relatives:

Wild relatives of tomato have been crucial in the improvement of cultivated tomato through classical breeding with regard to several traits such as pest and disease resistance, abiotic stress tolerance, and, to a much lesser extent, fruit quality. The Table 13.2 summarizes some crucial data in this connection obtained from the AVRDC—The World Vegetable Center (as of May 2013).

Use of disease and pest resistance genes of crop wild relatives in tomato breeding:

The incorporation of disease resistance through classical tomato breeding has been amply successful because of the availability of single major resistance genes, many of them with dominant inheritance. Virtually all significant resistance genes to tomato diseases were sourced from wild relatives. Rick and Chetalat (1995) listed a total of 42 major tomato diseases for which resistance genes have been identified in wild tomato relatives. Among those, S.*chilense*, S.*peruvianum*, S.*habrochaites* and S.*pimpinellifolium* (Table 13.2) were the richest sources. Hajjar and Hodgkin (2007) list a total of 55 traits, almost exclusively conferring pest and disease resistance, which were incorporated from wild relatives into released tomato varieties. Much of the research AVDRC focused on the introduction of resistance genes against late blight, bacterial wilt, and leaf curl (caused by Begomoviruses) into tomato varieties.

Table 13.2 Genetic stocks of Solanum section (*Lycopersicon*) (tomatoes) maintained by AVRDC with specific reference to wild species

Category	Description	Number of accessions
Wild species	S.*arcanum*	4
	S.*cheesmaniae*	17
	S.*chilense*	47
	S.*chmielewskii*	11
	S.*corneliomulleri*	11
	S.*galapagense*	17
	S.*habrochaites*	106
	S.*neorickii*	12
	S.*pennellii*	65
	S.*peruvianum*	133
	S.*pinpinellifolium*	323
Subtotal		746

However, as the main focus of this book would be on abiotic stresses and how crop wild relatives could contribute beneficial genes in this effort, the following discussion will be most pertinent in the context of this book.

Abiotic stress tolerance:

Plant response to environmental stress is highly influenced by environmental variation, and, in general, quantitatively inherited involving a multiple of genes. In addition, stress tolerance seems to be a stage-specific phenomenon, and tolerance at one stage of plant development is not necessarily correlated with tolerance at other stages (Foolad 2007). Nair and Khulbe (1990) have observed that there is differential response of wheat and barley genotypes to substrate-induced salinity under North Indian conditions. Thus, it is clear that the genes in a crop plant impart specific tolerance/resistance to abiotic stresses. Knowledge gained from the evaluation of developmental and physiological aspects of stress tolerance will facilitate a better understanding of its genetic basis and will aid the development of stress-tolerant cultivars. Therefore, screening for stress tolerance should be dissected into specific ontogenetic stages such as seed germination and emergence, seedling survival and initial plant growth, and vegetative growth and reproduction (Nair and Khulbe 1990).

Tolerance to heat stress:

All developmental stages of tomato are susceptible to heat stress (Peet et al. 1998). High temperatures have a negative impact on fruit production because of impaired pollen formation and development, fruit set, and fruit development (Peet et al. 1998). Sources of tolerance against heat stress can also be found in many wild relatives. *Solanum chilense* offers good prospects for raising levels of tolerance against high temperature (de la Pena et al. 2011).

Tolerance to salt stress:

Of the 14 billion ha of land on planet earth, 6.5 billion ha are salt affected. While genetic variability for salt tolerance traits is limited in cultivated tomato, in several wild relatives such as *Solanum pimpinellifolium, S.peruvianum, S.chilense, S.cheesemaniae, S.habrochaites, S.chmielewskii, S.esculentum var. cerasiforme, and S.pennellii* (Table 13.2) salt traits have been observed (Rao et al. 2013).

In *S.pimpinellifolium* a more effective detoxification mechanism and higher capacity to form lateral roots have been proposed as salt tolerance mechanisms (Sun et al. 2010). However, salt tolerance at one stage of plant development is genetically not correlated with salt tolerance at other developmental stages (Foolad and Lin 1997). Introgression of salt tolerance traits (genes) from distant wild relatives to cultivated *S.lycopersicum* is difficult because of crossability barriers and linkage drag.

The identification and use of salt tolerance traits in *S.pimpinellifolium* would be beneficial as it is the closest wild relative and readily crossable with *S.lycopersicum*. Many other important horticultural traits, including yield and disease resistance, have been identified in *S.pimpinellifolium* and used for the improvement of cultivated tomato (Foolad 2004). A subset of AVRDC's *S.pimpinellifolium* core collection has been evaluated to assess the effects of salt stress on physiological traits as well as yield-related traits with the aim of identifying potential accessions which could be used for salt tolerance breeding in tomato (Rao et al. 2013).

The identification of QTLs for salt stress at different developmental stages would facilitate simultaneous or sequential introgression of QTLs for tolerance and the development of tomato cultivars with improved salt tolerance at all important ontogenetic stages (Foolad 2007).

Tolerance to drought stress:

The tomato plant is sensitive to drought stress throughout its different developmental stages, from germination up to harvest. Genotypic variation for drought tolerance exists within the cultivated tomato and related to wild species. These are *S.cheesmaniae, S.chilense, S.pennellii, S.pimpinellifolium*, and *S.lycopersicum* var.*cerasiforme* (Foolad 2007). While most tomato cultivars are sensitive to drought stress at germination, sources of tolerance have been identified in *S.pennelli* and *S.pimpinellifolium* (Foolad 2007). Drought tolerance at the seed germination stage is a quantitative trait, and four QTLs have been identified on chromosomes 1, 8, 9, and 12. The alleles for drought tolerance were contributed by. *S.pimpinellifolium* at two loci, and *S.lycopersicum* at the other two loci (Foolad et al. 2003). Potential sources for drought tolerance during vegetative growth and reproduction have been identified in *S.chilense* and *S.pennellii*, mostly among accessions, native to dry habitats.

Different tolerance indices have been employed to characterize the physiological and/or genetic basis of drought tolerance in tomato (Foolad 2007). These include dry weight of shoot and root, root length, root morphology, leaf rolling, flower and fruit set, fruit weight, yield, water use efficiency, recovery after rewatering (irrigation), stomatal resistance, plant survival, leaf water and leaf osmotic potential, osmoregulation, oxidative damages, transpiration and photosynthetic rates, enzymatic activities, and pollen viability. This wide range of tolerance indices gives an indication of the complexity of this trait when selecting and breeding for drought tolerance. There is a very clear role to combine molecular tools with classical breeding techniques to achieve significant progress in developing cultivars with a high degree of drought tolerance.

Other abiotic stress tolerance:

Accessions of *Solanum esculentum var.cerasiforme, S.jugalandifolium* and *S.ochranthum* are tolerant against flooding (Robertson and Labate 2007), and tolerance against chilling injury has been reported for accessions of *S.habrochaites, S.chilense*, and S.lycopersicoides (Robertson and Labate 2007).

Molecular breeding to facilitate gene introgression from crop wild relatives into vegetable varieties:

Using crop wild relatives in vegetable breeding may present difficulties at different points such as trait discovery in crop wild relatives, trait introgression into cultivated genotypes, or during the recovery of the cultivated genotype through backcrossing of the hybrid to the recurrent parent. Molecular breeding methods can help to overcome these constraints and thus contribute to the successful use of crop wild relatives in vegetable breeding.

Gene and trait discovery in Crop Wild Relatives:

Crop Wild relatives (CWRs) are frequently used as sources for disease resistance/tolerance and for abiotic stresses, as cultivated genotypes lack such gene sources. Phenotypic screening of wild species for resistance against plant pathogens and insect pests is generally quite effective for such traits. But, the focus of this book is on abiotic stresses, in particular, high temperature leading to drought, a consequence of global warming.

Improvement of tolerance to abiotic stress through molecular breeding:

Introgression of tolerance traits from distant wild relatives to cultivated tomato has proven difficult because of crossability barriers and linkage drag. *Solanum pimpinellifolium* is the closest wild relative to cultivated tomato and easily crossable. *S.pimpinellifolium* is also a source of many useful genes for other traits such as yield and disease resistance (Foolad 2004). Recently, AVRDC has created a core collection of its *S.pimpinellifolium* variety comprising 322 accessions to make it more easily available to breeders (Rao et al. 2012). A subset of this collection was evaluated to assess the effects of salt stress on physiological traits as well as yield-related traits with the aim of identifying potential *S.pimpinellifolium* accessions which can be used for salt tolerance breeding in tomato (Rao et al. 2013).

13.2 Conclusion

Wild relatives of vegetable crops as sources of biotic and abiotic resistance/tolerance are likely to become more important in the days to come. This is because of the emerging climate scenario, and the consequent global warming, which will trigger higher demand for more vegetables, leading to greater investment in vegetable research, utilizing crop wild relatives, especially in a crop like tomato. Improved technologies, especially genomics-assisted breeding, are facilitating the introgression of favorable traits from wild species into cultigens. The conservation of genetic resources of wild relatives of vegetables and the full characterization of gene bank collections will be essential to provide the required new agronomic traits to breeding programs. Mobilization of the biodiversity available in the wild gene pool will allow cultigens to adapt to rapidly changing environmental conditions and boost agricultural production to ensure food and nutrition security.

Utilization of Crop Wild Relatives to improve horticultural traits of other vegetable crops:

Onion: *Allium roylei*, a wild relative of onion (*Allium cepa*), is considered a potential gene reservoir for onion breeding. A single dominant resistance gene against powdery mildew (*Peronospora destructor*) was introgressed from wild onion species *A.roylei* into cultivated onion (Kofoet et al. 1990). It took about two decades to get powdery mildew-resistant onion cultivars; the combination of the long generation time of onion, the genetic complexity of the crop, and linkage of resistance with a factor that is lethal when present at homozygous state in cultivated onion made the gene introgression a lengthy and difficult process (Scholten et al. 2007). However, there is no recorded instance where onion wild species have been used to impart abiotic stress resistance to climate change, among cultivated onion cultivars.

Brassica vegetables: The most economically important genus of the family *Brassicaceae* (= *Cruciferae*) is *Brassica*, which includes oilseed, forage, condiment, and vegetable crops. The main *Brassica* vegetable species is *Brassica oleracea* (kale, cabbage, broccoli, Brussels sprouts, cauliflower). *Brassica rapa* includes vegetable forms such as turnip, Chinese cabbage, and pak choi; and *Brassica juncea*, which is consumed as a vegetable, in Asia. Development of Brassica cultivars resistant to both biotic and abiotic stress and the production of lines with improved nutritional properties increasingly rely on introducing genes from exotic germplasm into elite breeding material (Scholze et al. 2003). However, sexual incompatibility barriers between different Brassica species make gene introgression from crop wild relatives into cultivars difficult. Somatic hybridization via protoplast fusion has been applied to overcome these barriers (Scholze et al. 2010).

Wild black mustard (*Brassica nigra*) represents one of the wild gene reservoirs to improve the resistance of cultivated varieties against several pathogens (Westman and Dickson 2000). Monosomic addition lines derived from backcrosses of somatic hybrids between *B.oleracea* var.*botyris* and *B.nigra* were developed as prematerials for resistance breeding in *B.oleracea* (Wang et al. 2011). Similarly, *Brassica fruticulosa* Cirillo (twiggy turnip) is a potential source for resistance of *Brassica* vegetables to cabbage aphid (*Brevicoryne brassicae*) (Pink et al. 2003) and cabbage root fly (*Delia radicum*). These interspecific allopolyploids could be used as bridge species to facilitate the use of crop wild relatives in *Brassicas* (Chen et al. 2011). However, there are no reported instances where such attempts have been made in combating climate change.

Lettuce: Most modern lettuce cultivars have been improved, thanks to traits sourced from wild relatives (Hajjar and Hodgkin 2007). All *Dm* (*Bremia lactucae*) resistant lines have their resistance traits derived from wild germplasm, and such cultivars have been regularly released since the 1980s (Crute 1992).

Lettuce RILs and advanced backcross inbred lines which carry chromosome segments of the wild lettuce species *L.serricola* or *L.saligna* in the background of *L.sativa* have been produced for QTL mapping (Zhang et al. 2007). Shelf-life related traits, as well as seed germination at elevated temperatures, were assessed in a RIL population (Zhang et al. 2007), while the advanced backcross lines were successfully used to map quantitative resistance genes to *Dm* (Jeuken et al. 2008). Both breeding strategies offer new possibilities for more durable *Dm* resistance in lettuce. However, no published work relates to the question of global warming and resistance/tolerance in lettuce.

References

Bai Y, Lindhout P (2007) Domestication and breeding of tomatoes: what have we gained and what can we gain in the future? Ann Bot 100:1085–1094

Barclay A (2004) Feral play: crop scientists use wide crosses to breed into cultivated rice varieties the hardiness of their wild kin. Rice Today 3:14–19

Battisti DS, Naylor RL (2009) Historical warnings of future food insecurity with unprecedented seasonal heat. Science 323(5911):240–244

Beachell HM, Khush GS, Aquino RC (1972) IRRI's international program. Rice breeding. IRRI, LosBanos, Philippines, pp 89–106

Beadle GW (1939) Teosinte and the origin of maize. J Hered 30:245–247

Brar DS, Khush GS (2003) Utilization of wild species of genus Oryza. In: Nanda JS, Sharma SD (eds) Monograph on genus Oryza. Science Publishers Inc, Enfield, Plymouth, pp 283–309

Burton-Freeman B (2011) Tomato consumption and heath emerging benefits. Am J Lifestyle Med 5(2):183–191

Caicedo A, Peralta I (2013) Basic information about tomatoes and tomato group. In: Liedl BE, Labate JA, Stommel JR, Slade A, Kole C (eds) Genetics, genomics and breeding of tomato. CRC Press, Taylor & Francis Group, Boca Raton, London, New York, pp 1–36

Chang TT (1970) Rice. In: Frankel OH, Bennett E (eds) Genetic resources in plants. Oxford and Edinburgh, UK, pp 267–272

Chaves MM, Oliveira MM (2004) Mechanisms underlying plant resilience to water deficits: prospects for water-saving agriculture. J Exp Bot 55(407):2365–2384

Chiarini FE, Moreno NC, Barboza GE, Bernardello G (2010) Karyotype characterization of Andean Solanoideae (Solanaceae). Caryologia 63(3):278–291

Cox S (2000) *I say tomayto, you say tomahto.* http://www.landscapeimagery.com/tomato.html. Accessed 5 Oct 2014

Crute IR (1992) From breeding to cloning (and back again?): a case study with lettuce downy mildew. Annu Rev Phytopathol 30:485–506

Darwin C (1839) Journal of researches: the voyage of the "Beagle". Originally Published as Voyages of the Adventure and Beagle, vol III–Darwin C (1839) Journal and remarks 1832–1836. Henry Colburn, London, UK

De la Pena RC, Ebert EW, Gniffke P, Hanson P, Symonds RC (2011) Genetic adjustment to changing climates: vegetables. In: Yadav SS, Redden RJ, Hatfield JL, Lotze-Campen H, Hall AE (eds) Crop adaptation to climate change, 1st edn. Wiley, Chicester, West Sussex, UK, pp 396–410 (Chap. 18)

Der Candolle A (1886) Origin of cultivated plants. Hafner Publishing Company, New York, USA (1959 reprint)

Ebert AW (2013). Ex situ conservation of plant genetic resources of major vegetables. In: Normah MN, Chin HF, Reed BM (eds) Conservation of tropical plant species. Springer Science + Business Media, New York, pp. 373–417. https://doi.org/10.1007/978-1-4614-3776-5-2 (Chap. 16)

Ellinger S, Ellinger J, Stehle P (2006) Tomatoes, tomato products and lycopene in the prevention and treatment of prostate cancer: do we have the evidence from intervention studies? Curr Opin Clin Nutr Metab Care 9(6):722–727

FAOSTAT (2013) http://faostat3.fao.org/home/index.html#VISUALIZE. Accessed 5 Oct 2014

Foolad MR (2004) Recent advances in genetics of salt tolerance in tomato. Plant Cell Tissue Organ Cult 76:101–119

Foolad MR (2007) Tolerance to abiotic stresses. In: Razdan MK, Mattoo AK (eds) Genetic improvement of Solanaceous crops. Tomato, vol 2. Science Publishers, Enfield, New Hampshire, USA, pp 521–590

Foolad MR, Lin GY (1997) Absence of a genetic relationship between salt tolerance during seed germination and vegetative growth in tomato. Plant Breed 115(4):363–367

Foolad MR, Zhang LP, Subbiah P (2003) Genetics of drought tolerance during seed germination in tomato: inheritance and QTL mapping. Genome 46(4):536–545

Chen J-P, Ge, X-H, Yao XC, Feng Y-H, Li ZY (2011) Synthesis and characterization of interspecific trigenomic hybrids and allohexaploids between three cultivated Brassica allotetraploids and wild species Brassica fruiticulosa. Afr J Biotechnol 10:12171–12176

Gould WA (1983) Tomato production, processing and quality evaluation, 2nd edn. AVI Publishing Company, Inc., Westport, Connecticut, USA, p 445

Hajjar R, Hodgkin T (2007) The use of wild relatives in crop improvement: a survey of developments over the last 20 years. Euphytica 156:1–13

Hanson PM, Yang R-Y, Wu J, Chen JT, Ledesma D, Tsou SCS (2004) Variation for antioxidant activity and antioxidants in tomato. J Am Soc Hortic Sci 129(5):704–711

Hijmans RJ, Jacobs M, Bamberg JB, Spooner DM (2003) Frost tolerance in wild potato species: assessing the predictivity of taxonomic, geographic, and ecological factors. Euphytica 130:47–59

Hoisington D, Khairallah M, Reeves T, Jean-Marcel R, Skomand B, Taba S, Warburton M (1999) Plant genetic resources: what can they contribute toward increased crop productivity. Proc Natl Acad Sci USA 96:5937–5943

Jeuken MJW, Pelgrom K, Stam P, Lindhout P (2008) Efficient QTL detection for nonhost resistance in wild lettuce: backcross inbred lies versus F2 population. Theor Appl Genet 116:845–857

Keatinge JDH, Yang R-Y, d'A.Hughes J, Easdown WJ, Holmer R (2011) The importance of vegetables in ensuring both food and nutritional security in attainment of the Millennium Development Goals. Food Secur 3:491–501

Kimber G (1993) Genomic relations in Triticum and the availability of alien germplasm. In: Damania A (ed) Biodiversity and wheat improvement. Wiley-Sayce, Chichester, UK, pp 9–16

Knapp S (2002) Tobacco to tomatoes: polygenetic perspective on fruit diversity in the Solanaceae. J Exp Bot 53:2001–2022

Kofoet A, Kik C, Wietsma WA, de Vries JN (1990) Inheritance of resistance to downy mildew (Peronospora destructor (Berk.Casp.) from Allium roylei Stearn in the backcross Allium cepa L. x (A.roylei x A.cepa). Plant Breed 105:144–149

Luckwill LC (1943) The genus Lycopersicon: an historical, biological and taxonomical survey of the wild and cultivated tomatoes. Aberdeen University Studies, vol 120, pp 1–44

Miller JC, Tanksley SD (1990) RFLP analysis of phylogenic relationships and genetic variation in the gene *Lycopersicon*. Theor Appl Genet 80:437–448

Mueller LA (2013) The tomato genome sequencing project. In: Liedl BE, Labate JA, Stommel JR, Slade A, Kole C (eds) Genetics, genomics and breeding of tomato. CRC Press, Taylor & Francis Group, Boca Raton, London, New York, pp 345–360

Mujeeb-Kazi A, Rosas V, Roldan S (1996) Conservation of genetic variation of Triticum tauschii (Coss.) Schmalh. (Aegilops suarrosa auct. Non L.) in synthetic hexaploid wheats (Tritcum turgidum L. s.lat. X T. tauschii 2n1/46x1/442, AABBDD) and its potential utilization for wheat improvement. Genet Resour Crop Evol 43:129–134

Nair KPP, Khulbe NC (1990) differential response of wheat and barley genotypes to substrate–induced salinity in North Indian conditions. Exp Agric 26(2):221–225 (Cambridge)

Peet MM, Sato S, Gardner RG (1998) Comparing heat stress effects on male-fertile and male-sterile tomatoes. Plant Cell Environ 21:225–231

Peralta IE, Knapp S, Spooner DM (2006) Nomenclature for wild and cultivated tomatoes. Tomato Genet Coop Rep 56:6–12

Peralta IE, Spooner DM, Knapp S (2008) Taxonomy of wild tomatoes and their relatives. Systematic botany monographs, vol 84 (*Solanum* sect. *Lycopersicoides* sect. *Juglandifolia* sect. *Lycopersicon: Solanaceae*). The American Society of Plant Taxonomists, p 186

Pink DAC, Kift NB, Ellis PR, McClement SJ, Lynn J, Tatchell GM (2003) Genetic control of resistance to the aphid *Brevicoryne brassicae* in the wild species *Brassica fruticulosa*. Plant Breed 122:24–29

Prasanna BM (2012) Diversity in global maize germplasm: characterization and utilization. J Biosci 37(5):843–855

Prescott-Allen C, Prescott-Allen R (1986) The first resource: wild species in the North American economy. Yale University Press, New Haven, CT, p 529

Rao ES, Kadirvel P, Symonds RC, Geethanjali S, Ebert AW (2012) Using SSR markers to map genetic diversity and population structure of *Solanum pimpinellifolium* for development of a core collection. Plant Genet Resour Charact Util. 10(1):38–48. https://doi.org/10.1007/S1479262111000955

Rao ES, Kadirvel P, Symonds RC, Ebert AW (2013) Relationship between survival and yield related trait in *Solanum pimpinellifolium* under salt stress. Euphytica 190(2):215–228. https://doi.org/10.1007/s10681-0120801-2

Rick CM (1995) Tomato: *Lycopersicon esculentum* (Solanaceae). In: Smartt J, Simmonds NW (eds) Evolution of crop plants, 2nd edn. Longman, Harlow, Essex, London, pp 452–457

Rick CM, Chetelat R (1995) Utilization of related wild species for tomato improvement. In: First international symposium on Solanacea for fresh market, vol 412, pp 21–38 (Acta Hortic)

Robertson LD, Labate JA (2007) Genetic resources of tomato (*Lycopersicon esculentum* Mill.) and wild relatives. In Razdan MK, Mattoo AK (eds) Genetic improvement of Solanaceous crops. Tomato, vol 2. Science Publishers Inc. Enfield, New Hampshire, USA, pp 25–75

Scholten OE, van Heusden AW, Khrustaleva LI (2007) The long and winding road leading to the successful introgression of downy mildew resistance into onion. Euphytica 156:345–353

Scholze P, Kramer R, Marthe F, Ryschka U, Klocke E, Schumann G (2003) Verbesserung der Krankheitsresistenz von Kohlgemuse: 2. Kohlhernie, Alternaria-und Phoma-Blattfleckenkrankheit. Gesunde Pflanzen 55(7):199–204

Scholze P, Kramer R, Ryschka U, Klocke E, Schuman G (2010) Somatic hybrids of vegetable brassicas as source for new resistance to fungal and virus diseases. Euphytica 176:1–14

Sears ER (1956) The transfer of leaf-rust resistance from Aegilops umbellulata to wheat. Brookhaven Symp Biol 9:1–22

Shannon MC (1997) Adaptation of plants to salinity. Adv Agron 60:75–120

Singh KB, Ocampo B (1997) Exploitation of wild Cicer species for yield improvement in chickpea. Theor Appl Genet 95:418–423

Strampelli N (1932) Origini, Svilluppi, Lavori e Resultati. Instituto Nazionale di Genetica per la cerealicoltura, Roma, Italian. (Origins, development, works and results. National Genetics Institute for Cereal Research, Rome, Italy)

Sun W, Xu X, Zhu H (2010) Comparative transcriptomic profiling of a salt-tolerant wild tomato species and a salt-sensitive tomato cultivar. Plant Cell Physiol 51(6):997–1006

Tanksley SD (2004) The genetic, developmental, and molecular bases of fruit size and shape variation in tomato. Plant Cell 16:181–189

The Tomato Genomic Consortium (2012) The tomato genome sequence provides insights into fleshy fruit evolution. Nature 485:635–641

USDA-ARS (2013) Family: Solanacae Juss. Nom. cons. National Genetic Resources Program. Germplasm Resources Information Network–(GRIN). National Germplasm Resources Laboratory, Beltsville, Maryland. http://www.ars-grin.gv/cgi-bin/npgs/html/family.pl?1043. Accessed 6 Nov 2014

Villareal RL, Sayre K, Banuelos O, Mujeeb-Kazi A (2001) Registration of four synthetic hexaploid wheat (Triticum turgidum/Aegilops taauchii) germplasm lines tolerant to waterlogging. Crop Sci 41:274

Wang G-X, Tang Y, Yan H (2011) Production and characterization of interspecific somatic hybrids between *Brassica oleracea* var. *botrytis* and *B.nigra* and their progenies for the selection of advanced pre-breeding materials. Plant Cell Rep 30:1811–1821

Westman AL, Dickson MH (2000) Disease reaction to *Alternaria brassicicola* and *Xanthomonas campestris* pv. *campestris* in *Brassica nigra* and other weedy crucifers. Cruciferae Newsl 22:87–88

Zhang FZ, Wagstaff C, Rae AM (2007) QTLs for shelf life in lettuce co-locate with those for leaf biophysical properties but not with those for leaf developmental traits. J Exp Bot 58:1433–1449

Chapter 14
The CWR of Minor Fruit Crops

The commonly used term "fruit" denotes edible botanical fruit, frutescens, (Latin word, meaning twiggy, bushy), and seed of wild and cultivated species of woody or herbal plants. A biologically important component of food, fruit is a rich source of nutrition as it contains a galaxy of vitamins, minerals, sugars and beneficial antioxidants, which give the human body many beneficial results, depending on which fruit one consumes. For humans, fruit, in the natural state, has been a rich source of nutrition and energy from time immemorial. Intended cultivation came much later. Paleoethno botanical data refer to the beginning of fruit cultivation in the second phase of plant domestication in Eurasia at the end of fourth millennium BC. The first evidence of cultivated grapes, olives, figs, and dates was found in excavations in the Mediterranean and near East. In colder parts of Europe and Asia, fruits were often collected as an admixture to the prevailing cereal and meat diet from paleolithic times. Apart from large temperate fruits such as apples, pears, plums, and cherries, there is a diversity of minor fruits available, such as strawberries, blueberries, blackberries, raspberries, cranberries, elderberries, cornelian cherries, rose hips, and various shrubby plums.

14.1 The Potential of Honeysuckle

There is no recorded research on the utility of wild relatives of minor fruit crops in breeding to evolve new types to combat global warming. Among them, the following brief description will be confined to the most important minor fruit crop, the honeysuckle (*Lonicera spp.* subsect. *Caerulea* Rehd.). This can serve as an excellent example of a boreal crop under the process of domestication, with a high potential for growing in new niches after climatic changes and, in addition, as an excellent health food. Honeysuckle is a promising fruit crop for a combination of highly positive advantages, such as, stable annual fruiting, earliness, and high biochemical parameters of the fruit. Boreal honeysuckles, characterized by high winter-hardiness, and adaptation to unfavorable environments, are used for breeding table cultivars with

© Springer Nature Switzerland AG 2019
K. P. Nair, *Combating Global Warming*, Springer Climate,
https://doi.org/10.1007/978-3-030-23037-1_14

relatively large fruit. Collection, selection, and introduction of promising new forms, and a breeding process based on crossing on the best forms from the large circumpolar distribution, can enhance the crop to be highly competitive with other commercial fruits. In addition, a high nutritional and medicinal value of the fruits and a very high organoleptic evaluation may enable honeysuckle to fill gaps in the health food market. The fruits of *L.caerulea* are a promising source of health beneficial substances which exhibit antiadherent, antioxidant, and chemoprotective properties.

14.1.1   The Other Fruits of Siberia

There are more than three hundred species of edible plants in the wild Siberian flora. Among these, there are sixty wild species of fruit plants, a greater part of which (about 45 species) are berries or small-fruit plants. Eleven species of local vegetation were introduced into cultivation (Gorbunov 1998). Small fruits, such as, black and red currant, (*Ribes*), gooseberry (*Grossularia*), bird cherry (*Cerasus avium*), raspberry (*Rubus idaeus*), blackberry (*Rubus fruticosus*), stone bramble (*Rubus saxatilis*), sea buckthorn (*Hippophae rhamnoides*), honeysuckle (*L.caerulea*), viburnum (*Viburnum lantana*), whortleberry (*Vaccinium* spp.), strawberry (*Fragaria*), cranberry (*Vaccinium oxycoccus*), blueberry (*Vaccinium myrtillus*) etc., can be found along river banks and lake shores, on islands, in new taiga amid forest steppe, in taiga and forest tundra. Siberian crab-apple (*Malus baccata* (L.) Borkh.) grows in flood plains and on mountain slopes of the Trans-Baikal region; penduncular almond (*Amygdalus pedunculata* Pall.) along the upper course of the Selenga River; Siberian apricot (*Armeniaca sibirica* (L.) Lam.) in Chita Province. West Siberia (Chelyabinsk and Tyumen Provinces) is the northeastern part of the area of European dwarf cherry (*Cerasus fruticosa* (Pall.) Borkh., which is also widespread in Europe (Malyshev and Peshkova 1994).

Siberian cold is extreme. Both soil and climatic conditions in Siberia are diverse and specific: winter temperatures go down to −40 to 45 °C, or even lower, while summers are relatively short and hot (up to 40–45 °C). Much snow falls in winter. Snow cover may settle down very late, when night temperature descends to −35 °C. Complex ecological conditions and the possibility of spontaneous hybridization between wild species over vast territories caused the development of numerous plant varieties, forms and hybrids, that now constitute a valuable natural genetic diversity of winter-hardy fruit and berry plants.

References

Gorbunov AB (1998). Introduction and breeding of food plants in the central siberian botanic garden. Sci Sib (31–32):28. (Original Article in Russian)

Malysev LI, Peshkova GA (1994) Berberidaceae—Grossulariaceae. Flora of Siberia, vol 7, 312 p. Nauka Publisher, Novosibirsk, Siberian Branch

Chapter 15
Ecosystem Services of Crop Wild Relatives

Agricultural sustainability will, increasingly, in the course of time, depend on using crop wild relatives to combat the ravages of global warming. If CWRs which may be critical to support worldwide food security comprise less than 0.26% of the world's flora (Maxted et al. 2012), and if they are "undervalued, underutilized and under threat" (Ford-Lloyd et al. 2011) especially from global climate change, land-use change, and other anthropogenic factors (Heywood 2011), then for how long will they be available to provide ecosystem services is a question that must grip everyone's attention concerned about the impact of climate change on global food security, knowing that their ancestors faced many climatic threats and survived them and contributed to increased likelihood of plastic adaptation of the current species (Bourou et al. 2012). The CWRs, like any other species, are being subjected to an increasing range of global threats, both biotic and abiotic, including invasion by alien invasive species (Ford-Lloyd et al. 2011). It has been estimated that 22% of plant species are threatened with extinction (www.kew.org/plants-at-risk), and up to 35% of the world's species could be on the path to climate-driven extinction (Minteer and Collins 2010). The rate of climate change is projected to be so rapid during the twenty-first century that many wild species, including CWRs, will be unable to adapt to the changed environment for survival (Mooney 2010). Enhanced pressures on natural resources, depletion of natural capital (steep soil fertility depletion, due to carbon depletion, thanks to the soil degrading fertilizer practices of the green revolution, the living example being what happened in Punjab State, the "cradle" of green revolution in India), and concerns about the impacts of environmental change have led to new research and policy agenda on the basis of the concept of ecosystem services (Maskell et al. 2013). Ecosystem services lie at the core of interactions among humans and ecosystems (Lavorel and Grigulis 2012) and they are functions of and provisions from ecosystems which are useful for and available to humans. Placing more emphasis on these services and their appropriate use should be useful for saving CWR species found outside protected areas, the desirability of which is widely acknowledged (Baidu-Forson et al. 2012).

© Springer Nature Switzerland AG 2019
K. P. Nair, *Combating Global Warming*, Springer Climate,
https://doi.org/10.1007/978-3-030-23037-1_15

15.1 Principles Involved

Conceptually, delivery of an ecosystem service increases with the level of intactness, complexity, and species richness of an ecosystem (Bommarco et al. 2013). If a "significant extinction of plant species is expected", when global average temperature increases by more than 3.5 °C (IPCC 2007), then the potential for loss of biodiversity, termination of evolutionary potential, and disruption of ecological services from CWRs must be taken more seriously (Dawson et al. 2011). Consequently, the capacity to enrich genetic diversity of CWRs becomes an essential component of their "ecological dynamics" to adapt them to current and future global climate change. Abundance and shifts in species distribution induced by climate change may affect biodiversity-related ecosystem services. Correlations between biodiversity and human population density suggest that people have long depended on biodiversity-related ecosystem services (Vitousek et al. 1997). Through complex interrelationships of biodiversity dynamics, ecosystem processes, and abiotic factors (Ayeni and Kambizi 2013), biodiversity influences the provision of ecosystem services through the strong links between biological species and a number of ecosystem processes, such as, pollination, climate regulation, and disease control, among others (Baidu-Forson et al. 2012). The components, for instance, genes, species or traits, and attributes, for instance, amount, variability, or composition of biodiversity which are necessary or desirable to retain any specific ecosystem service will vary according to the service being considered and the ecosystem processes on which it depends (Mace et al. 2012). Diversifying climatic, edaphic, and biotic natural selection is a major evolutionary force driving trait differentiation at single and multilocus levels in many CWRs (Loreau and de Mazancourt 2013). The resulting polymorphisms found at different levels of genome organization may have constituted a regulatory adaptation of CWRs to climatically fluctuating environments, both at the macro and micro environmental and edaphic levels.

Despite their importance as a critical resource for the future well being of humans, CWRs have not been adequately recognized in the discussions pertaining to ecosystem services (Ford-Llyod et al. 2011). Along with their domesticated progenies of food grains, legumes, horticultural items, such as, fruits and vegetables, medicinal and aromatic spices, such as turmeric and ginger, (Nair 2013), ornamental crops, forage, and pasture crops, root crops, and agroforestry, CWRs provided, for millennia, multiple ecosystem services to humans (Heywood 2011). In the future decades to come, some CWRs species may face extreme selection pressure and alterations to their functional composition because of the magnitude and rapidity of climate change, and, may not be able to adequately adapt to selection pressures (Craine et al. 2013). Of particular concern is the impact of the global warming trend in tropical and subtropical ecosystems (Arpaia et al. 2012) and recurring drought in dryland ecosystems, where several centers of origin and diversity of CWRs exist, as discussed in earlier chapters.

It is vital that continued survival and evolution of CWRs, providing genetic diversity for breeding, development of new crop varieties, and provisioning of ecosystem services, is ensured. It is imperative to distinguish between a conservation unit, which depends on the conservation goals, and a service-providing unit, which offers, or might offer in the future, a recognized ecosystem service at some spatiotemporal scale (Luck et al. 2003). Nevertheless, one must ensure that CWRs populations have sufficient evolutionary potential to respond to future climate change patterns. It is quite probable that CWR germplasm will need to be moved around the world, more than ever, to facilitate the process of agricultural adaptation, in response to changing climate patterns. The renewed interest in the genetics and breeding of CWRs to provide new ecosystem services (Huang et al. 2009) highlights the growing importance of these species as sources of quantitative traits. A comprehensive list of major plant families of CWRs and Wild Under-Utilized Species (WUS) in use or of potential use in all agricultural crops is an urgent need (Khoury et al. 2013). Unfortunately, a major obstacle preventing the application of CWRs and WUS, as a regular tool for crop improvement is a total lack of adequate knowledge about the genetics of traits of interest in the wild germplasm (Aerts et al. 2012).

15.2 The Impact of Climate Change on Biodiversity of CWR

There is divergent opinion, among conservationists, whether focusing on ecosystems services might lead to loss of CWR biodiversity or not. Biodiversity is linked to climate change, and this includes the biodiversity of CWRs. As a consequence of rapidly changing impacts from pathogen-exposure, because of climate change, CWRs and WUS may experience genetic erosion or even extinction. A relevant question is, how will climate change, together with other environmental stressors, alter the distribution and prevalence of diseases of wild species (Sutherland 2006)? Climate change is already linked to a range of biotic and abiotic impacts at the species level, including physiological, phenological, and distributional changes (Minteer and Collins 2010). Many CWRs are found in disturbed, preclimax, plant communities, which are the habitats which are likely to be subjected to increasing levels of anthropogenic change (Maxted et al. 2012). It is worth noting, in this context, that physiology, life history, and other genetic-based elements of adaptive capacity may differ substantially for CWR populations at the core of the species distribution versus those at the boundaries, implying that such populations may have different responses to similar climatic change (Nevo 1998). These varied responses can be attributed to either convergent evolution because of lower differentiation of a quantitative trait compared with neutral loci, suggesting that similar phenotypic values are favored in different environments, or to adaptive differentiation of a quantitative trait as compared with neutral loci (Jump et al. 2008).

An important aspect in the assessment of climate change impact on the distribution of CWRs is their ability for dispersal. Some CWR populations conserved in vulnerable locations are unlikely to be able to adapt sufficiently quickly to climate change;

other populations may go through genetic bottleneck, thus reducing the long-term variability to sustain their diversity in situ. If, in the near future, traditional in situ conservation methods will be inadequate to save threatened species from extinction because of climate change, do we need then novel ecosystems to save such stressed populations, without consideration of potential invasion risk of the relocated species in their new habitats (Minteer and Collins)? Contrarily, if migration of species is low and natural selection is adequately strong, then local adaptation may lead to differentiation between CWR populations over time. Genetic drift may also result in genetic differentiation especially in self-pollinated plants (Nevo 1998). These differential responses emanate either from convergent evolution because of lower differentiation of a quantitative trait compared with neutral loci, suggesting that similar phenotypic values are favored in different environments, or to adaptive differentiation of a quantitative trait as compared to neutral loci (Jump et al. 2008).

CWR populations contain considerable genetic diversity and include allelic variants which can enhance reproductive success under a wide range of environmental stresses and climate regimes (Assmann 2013). Drought, high temperature, frost, acidity, and salinity, with their attendant toxic mineral ions and mineral stress, are expected to become more unpredictable and severe, in the coming years, which will persist as major abiotic stressors limiting crop productivity. Adaptation to climate change will require new traits which might not be available in the germplasm of cultivated crops; this will further increase the importance and demand for CWRs (Cobben et al. 2013). By directed genetic introgression from CWRs, for drought, acidity, salinity and temperature tolerance (frost and heat), the gene pool of cultivated crops can be substantially enhanced (Redden 2013). The above-mentioned are the most important abiotic stress factors which are likely to pose as very serious challenges of climate change, especially at the plant reproductive stage. The abiotic stress factors will alter the important mineral composition of soil, required for optimal plant nutrition, and also at the plant root surface, and will have dramatic negative impact on crop productivity, as has been amply demonstrated by Nair (2010, 2013). Climate change will exacerbate these problems because as temperatures continue to rise, because of global warming, evaporative demand also will increase as a result of the decrease in atmospheric water potential with increasing temperature (Assmann 2013). "Ionomics" analysis was successful in identifying genetic variation in CWR phenotypes associated with salinity and with both beneficial mineral nutrients and reduced toxic mineral ions, yet, mineral stress remains as the missing link in our understanding of how climate stress will affect plant productivity. Allelic variation in some model CWRs for phenotypic traits relevant to the future climate and anthropogenic environment shows that considerable genetic diversity exists for such traits (Assmann 2013).

On account of the magnitude and acceleration of climate change, some CWRs are quite likely to face extreme selection pressure in the decades to come. Of special significance in this context is the extreme warming trends in tropical and subtropical ecosystems (Arpaia et al. 2012), and, recurring drought episodes in dryland ecosystems (Bernard et al. 2012), as in India, for example, where several centers of origin of CWR exist. Recurring drought, in particular, may alter the functional composition and impact ecosystem services (Craine et al. 2013). A contemporary example is ecosystems harboring CWRs of fruit trees in Central Asia which are associated with inherently different levels of productivity and disturbance. These ecosystems may respond differently to the same anthropogenic stressors (Maskell et al. 2013). The FAO (2006) has reported that several species of CWR of fruit trees in Central Asia are facing threats from habitat destruction, overgrazing, overharvesting, and the increasing impact of global climate change. Some CWRs, for instance wild apple, are more diverse and phenotypically plastic than others, for instance, wild almond, which was included in the "Red List" (read earlier discussion). Such large phenotypic plasticity can shift adaptive capacity under a rapidly changing climate pattern without relying on genetic diversity (Walther 2010). Additionally, transgressive segregates of natural hybrid population between domesticated fruit trees and their CWRs may allow these hybrids to survive in habitats which are more extreme than those of either parent (Jansky et al. 2013).

15.3 Ecosystem Services of CWRs Functional Biodiversity

For short-term ecosystem resource dynamics and long-term ecosystem stability, functional diversity of CWR traits—the value and range of CWR traits rather than just species numbers—is vital as it increases positive interactions or complementarity functions within an ecosystem, whereas genetic diversity, through genotypic complementarities, can buffer against extreme climatic events, thus replacing the role of species diversity (Hajjar et al. 2008). Genetic diversity may be more likely to affect the resistance of ecosystems to perturbation than to affect ecosystem processes under normal conditions; therefore, phenotypic variance, rather than a species diversity metric, may be a more appropriate measure of diversity in attempting to relate diversity to ecosystem functioning and dynamics (Hajjar et al. 2008).

Within a CWR, one can identify which genetic variance (s) is more beneficial than another. For the persistence of a CWR species or population in its natural environment, one must, however, assume that all genetic variation is valuable per se (Jump et al. 2008). Contrarily, cryptic diversity, that is, genetic diversity within a species, is the raw material on which evolution acts and may provide "genetic refugia", which harbor the exclusive genetic solutions to deal with climate change. The latter may result not only in the extinction of a species, but also, in a significant loss of cryptic diversity (Assmann 2013). A loss of cryptic diversity may result in shorter useful lives, for instance, of resistance genes to biotic and abiotic stresses (Cheatham et al. 2009). Natural selection for survival with better seed set on the more tolerant plants could result in higher genetic variation for tolerance to abiotic stresses in CWRs than in the primary gene pool of most domesticated crops to survive in the predicted extremes of climate change, and relies, in part, on CWR's with untapped genetic variation for abiotic and biotic stress tolerance. This "survivalomics" strategy (Redden 2013) was suggested as a means to expand available domesticated gene pools to assist crops to survive in the predicted extremes of climate change, and relies, in part, on CWRs with untapped genetic variation for biotic and abiotic stress tolerances. Suitability of parental CWR germplasm for introgression of abiotic stress tolerance can be based on the species survival in extreme habitats (Redden 2013). Through the "survivalomics" approach, geneticists and plant breeders can locate sources of tolerance against biotic and abiotic stresses; however, it will be crucial to optimize genomic tools to successfully incorporate such traits in new crop varieties. This approach may have enormous potential to provide insights into functional allelic variation which can be harnessed to improve crop productivity and possibly ameliorate climate-change effects on crop yield (Assmann 2013). There is new evidence to demonstrate the relevance of functional plant traits as strong candidates to quantify ecosystem service delivery, given their effects on underlying ecosystem processes (Lavorel and Grigulis 2012). Additionally, the opportunistic weedy character of some CWRs, for example, the model plant *Arabidopsis thaliana*, may indicate underlying genetic variation in plasticity-related genes, which may be capitalized on for crop improvement (Assmann 2013).

An important aspect to understand is: Have there been vast changes in today's fruit trees compared to their ancestors to tap the potential of using their CWR and WUS? The answer to this important question is an emphatic "no". To substantiate, it is noted that many fruit trees of today have diverged but very little from their ancestors. In apple (*Malus* spp.), for instance, the present-day cultivars are almost not distinct from the selections obtained from wild stands in Central Asia (FAO 2006). Also, the high diversity of almond (*Amygdalus* spp.) includes CWR segregates that are very close to domesticated almond. This would explain why the ecological adaptation of the classic Mediterranean fruits has not exceeded the requirements of their wild ancestors (Janick 2005). Unlike CWRs, there is little information on how variation within WUS plants may respond to selection pressures through management, soil, or climate, and if such variation has ecological significance in influencing food webs or crop-WUS competition (Begg et al. 2012). Some WUS, for example, *Capsella bursa-pastoris*, mediate essential ecological processes and provide ecosystem services in managed ecosystems. Such species generally function better with increasing diversity levels (Bommarco et al. 2013). The above facts would unequivocally point to the inescapable conclusion that tapping the CWR and WUS, to combat global warming, will fetch better results in food crops like wheat and rice, the mainstay of humankind, than fruits, which only form supplementary diet. This conclusion has great significance in policy planning for the future.

References

Aerts R, Berecha G, Gijbles P (2012) Genetic variation and risks of introgression in the wild Coffea arabica gene pool in south-western Ethiopian montane rainforest. Evol Appl 6:243–252. https://doi.org/10.1111/j.-1752-4571.2012.00285.x

Arpaia S, Battafarano R, Chen L-Y, Devos Y, Di Leo GM, Lu BR (2012) Assessment of transgene flow in tomato and potential effects of genetically modified tomato expressing Cry3Bb1 toxins on bumblebee feeding behavior. Ann Appl Biol 161:151–160

Assmann SM (2013) Natural variation in abiotic stress and climate change response in Arabidopsis: implications for twenty-first-century agriculture. Int J Plant Sci 174(1):3–26

Ayeni O, Kambizi L (2013) The interrelation of biodiversity dynamics, ecosystem processes and abiotic factors: a review. Afr J Basic Appl Sci 5(1):1–7

Baidu-Forson JJ, Hodgkin T, Jones M (2012) Introduction to special issue on agricultural biodiversity, ecosystems and environment linkages in Africa. Agric Ecosyst Environ 157:1–4

Begg GS, Wishart J, Young MW, Squire GR, Lannetta PRM (2012) Genetic structure among arable populations of Capsella bursa-pastoris is linked to functional traits and in-field conditions. Ecogeography 35:446–457

Bernard G, Anthelme F, Baali-Cherif D (2012) The Laperrine's olive tree (Oleaceae): a wild genetic resource of the cultivated olive and a model—species for studying the biography of the Saharan Mountains. Acta Bot Gallica 195(3):319–328

Bommarco R, Kleijn D, Potts SG (2013) Ecological intensification: harnessing ecosystem services for food security. Trends Ecol Evol 28:230–238

Bourou S, Bowe C, Diouf M, Van Damme P (2012) Ecological and human impacts on stand density and distribution of tamarind (Tamarindus indica L.) in Senegal. Afr J Ecol 50:253–265

Cheatham MR, Rouse MN, Esker PD (2009) Beyond yield: plant disease in the context of ecosystem services. Phytopathology 99(11):1228–1236. https://doi.org/10.1094/PHYTO-99-11-1228

Cobben MMP, van Treuren R, van Hintum TJL (2013) Climate change and crop wild relatives: can species track their suitable environment, and do what do they lose in the process? Plant Genet Resour 11(3):234–237. https://doi.org/10.1017/S1479262113000087

Craine JM, Ocheltree TW, Nippert JB (2013) Global diversity of drought tolerance and grassland climate-change resilience. Nat Clim Change 3:63–67. https://doi.org/10.1038/NCLMATE1634

Dawson TP, Jackson ST, House JI, Prentice IC, Mace GM (2011) Beyond predictions: biodiversity conservation in a changing climate. Science 332:53–58

FAO (2006) People, forests and trees of West and Central Asia. Outlook for 2020. FAO Forestry Paper No. 152. FAO, Rome, Italy

Ford-Lloyd BV, Schmidt M, Armstrong SJ (2011) Crop wild relatives—undervalued, underutilized and under threat? Bioscience 61:559–565

Hajjar R, Jarvis DI, Gemmil-Herren B (2008) The utility of crop genetic diversity in maintaining ecosystem services. Agric Ecosyst Environ 123:261–270

Heywood VH (2011) Ethnopharmacology, food production, nutrition and biodiversity conservation: towards a sustainable future of indigenous peoples. J Ethnopharmacol 137:1–15

Huang L, Brooks S, Li W, Fellers J, Nelson JC, Gill B (2009) Evolution of new disease specificity at a simple resistance locus in a crop–weed complex; reconstruction of the Lr21 gene in wheat. Genetics 182:595–602

IPCC (2007) In: Parry ML, Canziani OF, Palutikof JP, Van der Linden PJ, Hanson PM, Yang R-Y, Wu J, Chen JT, Ledesma D, Tsou SCS (2004) Variation for antioxidant activity and antioxidants in tomato. J Am Soc Hortic Sci 129 (5):704–711

Janick J (2005) The origin of fruits, fruit growing, and fruit breeding. Plant Breed. Rev. 25:255–320

Jansky SH, Dempewolf H, Camadro EL (2013) A case for crop wild relative preservation and use in potato. Crop Sci 53:746–754

Jump AS, Marchant R, Penuelas J (2008) Environmental change and the option value of genetic diversity. Trends Plant Sci 14:51–58

Khoury CK, Greene S, Wiersema J, Maxted N, Jarvis A, Struik PC (2013) An inventory of crop wild relatives of the United States. Crop Sci 53:1496–1508

Lavorel S, Grigulis K (2012) How fundamental plant functional trait relationships scale-up to trade offs and synergies in ecosystem services. J Ecol 100:128–140

Loreau M, de Mazancourt C (2013) Biodiversity and ecosystem stability: a synthesis of underlying mechanisms. Ecol Lett 16:106–115

Luck GW, Daily GC, Ehrlich PR (2003) Population diversity and ecosystem services. Trends Ecol Evol 18(7):331–336

Mace GM, Norris K, Fitter AH (2012) Biodiversity and ecosystem services: a multilayered relationship. Trends Ecol Evol 27(1):19–26

Maskell LC, Crowe A, Dunbar MJ (2013) Exploring the ecological constraints to multiple ecosystems service delivery and biodiversity. J Appl Ecol 50:561–571

Maxted N, Kell S, Ford-Llyod-B V, Dulloo ME, Toledo A (2012) Toward the systematic conservation
of global crop wild relative diversity. Crop Sci 52(2):774–785

Minteer BA, Collins JP (2010) Move it or lose it? The ecological ethics of relocating species under
climate change. Ecol Appl 20(7):1801–1804

Mooney HA (2010) The ecosystem—service chain and the biological diversity crisis. Philos Trans
R Soc B 365:31–39

Nair KPP (2010) "The Nutrient Buffer Power Concept"—a revolutionary soil management technique. In: Proceedings of the 19th world congress of soil science, Brisbane, Australia

Nair KPP (2013) The buffer power concept and its relevance in African and Asian soils. Adv Agron
121:416–516

Nevo E (1998) Genetic diversity in wild cereals; regional and local studies and their bearing on
conservation ex-situ and in-situ. Genet Resour Crop Evol 45:355–370

Redden R (2013) New approaches for crop genetic adaptation to the abiotic stresses predicted with
climate change. Agronomy 3:419–432

Sutherland WJ (2006) Predicting the ecological consequences of environmental change: a review
of the methods. J Appl Ecol 43:599–616

Vitousek PM, Mooney HA, Lubchenko J, Melilo JM (1997) Human domination of earth's ecosystems. Science 277:494–498

Walther GR (2010) Community and ecosystem responses to recent climate change. Philos Trans R
Soc B 365:2019–2024

Chapter 16
Predictive Characterization of CWRs

Within the CWR population diversity, the principal components are richness, size, spatial distribution, and differentiation (Luck et al. 2003). The distribution of population sizes determines whether a CWR is characterized by a single large population, many small populations, or several similarly sized populations. Population distribution is a measure of the spatial aggregation of populations and how this can affect the delivery of ecosystem services, including genes for qualitative, for example disease resistance, and quantitative, for example drought, frost and/or acidity, salinity tolerance, biotic and abiotic stresses, respectively. Spatial analysis was suggested as a means of predictive characterization to predict which CWR germplasm might have desired traits (Maxted et al. 2012). For example, CWR populations found in hot spots of pests and diseases are likely to have evolved resistances over time to these biotic stresses.

Climate change impacts ecosystem in such a complex manner that effective prediction requires consideration and modeling of a wide range of interacting factors (Galic et al. 2012). While modeling ecosystem services delivery from CWRs, trait responses to environmental and management drivers and trait effects on ecosystem properties underlying the provisioning of ecosystem service are needed for predictive characterization purposes (Lavorel and Grigulis 2012). Predictive characterization of CWRs at the reproductive stage should be augmented by characterization at the vegetative stage. Vegetative traits are often correlated with fitness, that is, reproductive success, seed or propagule production. Therefore, characterization of CWRs, at the vegetative stage constitutes an essential step in germplasm enhancement (Assmann 2013). It is an indispensable pre breeding phase and would lead to the development of genotypes and cultivars tolerant to abiotic and biotic stresses, respectively, such as, heat, cold, drought, salinity and pest, and disease (Campbell et al. 2008).

There is an increasing interest in geographic information system (GIS) in the predictive characterization of CWR. In addition to modeling approaches, GIS analyses should facilitate the identification and characterization of population and species of CWRs and other useful WUS, most at risk from climate change (Galic et al. 2012). A few GIS-based investigations have been made to compare in situ diversity of CWRs with diversity which has been sampled and conserved either in situ or ex situ

© Springer Nature Switzerland AG 2019
K. P. Nair, *Combating Global Warming*, Springer Climate,
https://doi.org/10.1007/978-3-030-23037-1_16

(Maxted et al. 2012). Such investigations indicated that *Hordeum* spp. with the highest priority for in situ conservation are those with the highest potential socioeconomic value including *Hordeum brevisubulatum, Hordeum bulbosum, Hordeum chilense, Hordeum jubatum, and Hordeum vulgare* subsp. *spontaneum.* These species provide in situ ecosystem services and have good prospects of utilization in barley breeding for qualitative and quantitative traits. Additional evidence comes from long-term spatiotemporal investigations conducted on major and minor populations of wild emmer wheat *(Triticum dicoccoides)* in the Fertile Crescent (Nevo 1998). Population fragmentation, in this and other CWRs and its impact on population structure and differentiation, most probably resulted from past climatic conditions which have been exacerbated by human action (Chair et al. 2011).

References

Assmann SM (2013) Natural variation in abiotic stress and climate change response in Arabidopsis: implications for twenty-first-century agriculture. Int J Plant Sci 174(1):3–26

Campbell A, Chenery A, Coad L (2008) The Linkages between biodiversity and climate change mitigation. UNEP World Conservation Monitoring Centre, Montreal, Canada

Chair H, Duroy PO, Curby P, Sinsin B, Phams JL (2011) Impact of past climatic and recent anthropogenic factors on wild yam genetic diversity. Mol Ecol 20:1612–1623

Galic N, Schmolke A, Forbes VE, Baveco H, van den Brink P (2012) The role of ecological models in linking ecological risk assessment to ecosystem services in agroecosystems. Sci Total Environ 415:93–100

Lavorel S, Grigulis K (2012) How fundamental plant functional trait relationships scale-up to trade offs and synergies in ecosystem services. J Ecol 100:128–140

Luck GW, Daily GC, Ehrlich PR (2003) Population diversity and ecosystem services. Trends Ecol Evol 18(7):331–336

Maxted N, Kell S, Ford-Llyod-B V, Dulloo ME, Toledo A (2012) Toward the systematic conservation of global crop wild relative diversity. Crop Sci 52(2):774–785

Nevo E (1998) Genetic diversity in wild cereals; regional and local studies and their bearing on conservation ex-situ and in-situ. Genet Resour Crop Evol 45:355–370

Chapter 17
CWR and Pre Breeding in the Context of the International Treaty of Plant Genetic Resources for Food and Agriculture (PGRFA)

17.1 Conservation and Management of Plant Genetic Resources for Food and Agriculture (PGRFA) in Wild Ecosystems

On the basis of a broad definition of CWR, as any taxon belonging to the same genus as a crop, it has been estimated that there are 50–60,000 CWR species worldwide. Of these, approximately 700 are considered of highest priority and are the species which comprise the primary and secondary gene pools of the world's most important food crops, many of which are included in the Annex 1 of the International Treaty on Plant Genetic Resources for Food and Agriculture (ITPGRFA) (Maxted and Kell 2009). The objectives of the treaty are the conservation and sustainable use of PGRFA and the fair and equitable sharing of the benefits arising out of their use, in harmony with the Convention on Biological Diversity, for sustainable agriculture and food security. Although in the ITPGRFA there is only a reference to CWRs in the article 5.1 (d), the impact of the management of those crops is very relevant for the conservation and sustainable use of PGRFA and their management in in situ and on-farm conditions. PGRFA is defined as "any genetic material of plant origin of actual or potential value for food and agriculture", which includes CWR. The CWR is a genetic portfolio which has a strategic use for plant breeders. Such germplasm will ultimately contribute to enhance adaptation of agriculture to climate change.

Information generated by research programs is also very relevant to inform policy makers, in particular, within the treaty. The knowledge arising from some projects, such as, the "Global Crop Diversity Trust Project on Crop Wild Relatives" is very valuable and provides updated information on how the PGRFA is essential to food security, and how countries depend largely on plant diversity originating elsewhere. These were the two criteria which were drawn upon to create the Multilateral System of the Treaty, so the information and knowledge generated should be taken into account for a future revision of the criteria to provide facilitated access to CWR germplasm.

© Springer Nature Switzerland AG 2019
K. P. Nair, *Combating Global Warming*, Springer Climate,
https://doi.org/10.1007/978-3-030-23037-1_17

Reference

Maxted N, Kell SP (2009) Establishment of a global network for the in situ conservation of crop wild relatives: status and needs. FAO Commission on Genetic Resources for Food and Agriculture, Rome, Italy, p 266

Chapter 18
Pre Breeding by Utilizing CWR

It has been the beaten track, for over decades now, by those involved in plant breeding, to use the limited number of parental lines to develop varieties of major food crops like rice and wheat. The current unfolding scenario on the climate front calls for a new thinking, because, along with dramatic climate change, new strains of virulent pests and diseases have also surfaced. The principal reason why, some of the "elite" crop varieties, created during the green revolution phase, have miserably failed, is because of their narrow genetic base. For some time, these elite varieties, whether they be of rice or wheat, thrived, but, in no time, the pests got smarter than them. This scenario rightly evokes the specter of the recurrence of historically devastating crop failures which resulted from particularly virulent pests and diseases. Examples are the late blight of potato, of the 19th century in Ireland, the rice stem rust of 1942 in India, the southern leaf blight of corn of 1970 in USA. More importantly, the "dwarf miracle" rice varieties from the International Rice Research Institute (IRRI) in the Philippines, and, wheat from the Center for Maize and Wheat Research (CIMMYT) in Mexico, which ushered the green revolution in south Asia, in particular in India, soon fell victim to diseases like blast, rust, and insect pests like brown plant hopper (BPH). Entire fields were destroyed by these diseases and pests in India. Coupled with the soil degradation, a result of unbridled use of chemical fertilizers and pesticides, the green revolution fell on it's face by early eighties in India, in a span of about a decade and a half, having been, thoughtlessly, pushed forward in mid sixties, when India had other options open for it to execute to meet the food needs of the country. The experience of other neighboring countries like Pakistan, Bangladesh, Philippines, Indonesia, or even China, is no different. The highly soil extractive farming took a heavy toll on soil resources of South Asia. Though, for a decade and slightly more, the country produced some excess foodgrains, the environmental costs the nation paid were huge. The degraded soils, the barren fields, the nitrate loaded ground water making it non potable and the spreading cancer, due to indiscriminate use of pesticides and herbicides, in parts of Punjab, like the Gurdaspur district, became the living testimony to a fallen green revolution. And, the devastating environmental fallout surfaced and spread to other parts of India, where the same model of the green revolution was followed. The narrow genetic base of the "new" and "miracle" crop varieties of

© Springer Nature Switzerland AG 2019
K. P. Nair, *Combating Global Warming*, Springer Climate,
https://doi.org/10.1007/978-3-030-23037-1_18

Table 18.1 Crop Wild Relatives in the First Cycle of the BSF Projects under ITPGRFA

Scope of the Fund: The fund seeks to accelerate the conservation and use of plant genetic resources on a global scale through technology transfer, capacity building, high-impact projects, and innovative partnerships involving farmers, plant breeders, civil society, and other stake holders

Projects in Cuba, Egypt, Morocco, Nicaragua, Senegal, Tanzania, and Uruguay carried out surveys and established crop inventories, collecting a total of 355 accessions of 12 different traditional and locally adapted crop varieties, as well as, wild relatives of crops (beans, citrus, cucurbits, finger millet, lablab beans, maize, *Plectranthus* sp., tomato, potato, wheat, sorghum, and yams) for research purposes

The Nicaragua and Uruguay projects both had a strong focus on CWRs. In Nicaragua, Teosinte populations naturally occurring in the Genetic Resources Reserve of Apacunca were mapped and collected together with six wild tomato relatives, for characterization and ex situ conservation. In Uruguay, 19 accessions of wild potato relatives were collected in 9 of the countries' administrative departments, with the aim of characterizing them regarding their resistance to bacterial wilt (*Ralstonia solanacearum* (Smith) Yabuuchi), and subsequently breeding resistance traits into cultivated potato varieties

rice and wheat showed. It is in this context that one must draw inspiration from the wide possibilities of utilizing CWRs of these crops to combat global warming and pestilence. The above Table 18.1 shows how risks of food security because of climate change could be reduced by appropriate conservation and use of genetic resources and links this to activities under the treaty to conserve and make available new genetic material.

Pre breeding is a vital step in the link between conservation and use of plant genetic resources. Limited funding, unfortunately, during the past years, has been a huge constraint in public funding for more basic research, which has seriously limited current work in pre breeding, in particular, national agricultural problems. The farmers require a diverse portfolio of well-adapted crop varieties to have resilient and varied farming systems needed to feed the burgeoning global population. In reality, just the opposite is happening. There is ample evidence to show that the cultivars of major crops are increasingly sharing greater genetic similarities than has hitherto been the case (Mercer et al. 2010). The narrow genetic base, which is a consequence of this, makes them victims of both biotic and abiotic stresses, and, stands in the way to breach yield barriers.

A spectrum of partnerships has been established through these pre breeding projects. While some are traditional like the Consultative Group for International Agricultural Research (CGIAR) center—National Agricultural Research Systems (NARS) center partnerships, other projects explore different settings, including North-South or South-South models of partnerships of technology transfer. There could be valuable practical lessons learned from such projects which would be useful for others, including the partners involved in developing the Platform for Co-Development and Transfer of Technology under the Treaty. The use of Standard Material Transfer Agreement or specific licensing agreements to exchange pre breeding material, even obtained from CWR, is the key to facilitate the access to those crops, and it can be an easy solution to monitor the distribution of PGRFA.

Table 18.2 Crop Wild Relatives in the Second Cycle of the BSF Projects under the ITPGRFA	Scope of the Fund: The Fund seeks to encourage participatory and science-based formulation of a strategic plan of action to strengthen conservation of plant genetic resources and their enhanced use in adapting to climate change in Mesoamerica The country-wise details are given below: Costa Rica: Participatory and science-based formulation of a strategic action plan to strengthen the conservation of plant genetic resources and their enhanced utility to adapt to climate change, and, thereby combat global warming, in Mesoamerica. A diagnosis of the status of PGRFA in the region has been generated focusing on 10 Mesoamerican crops and their wild relatives, gene pools of corn (*Zea mays*), beans (*Phaseolus* spp.), cassava (*Manihot*), sweet potato (*Ipomoea*), cucurbits (*Cucurbita*), amaranth (*Amaranthus*), peppers (*Capsicum*), avocado (*Persea*) and forage (*Tripsacum*) [a] Jordan/Iran: Use of genetic resources to establish a multi-country program of evolutionary—participatory plant breeding. A collecting mission has been organized in Iran and Jordan which focused, in particular, on the wild progenitors of wheat and barley, which are still under-represented in the existing collections and which hold promise as source of genes for adaptation to the harsher climatic conditions expected in the future Malawi: Increased genetic diversity of target crops conserved in the national, regional, and international gene banks and on farm. Additional 200 accessions of sorghum, pearl millet, finger millet, yams, and cowpea germplasm including their CWRs not available from the gene bank were collected and included in the evaluation trials

[a]http://www.biodversityinternational.org/uploads/tx_news/SAPM_2014_2024_1725pdf 18 pp

18.1 ITPGRFA Benefit-Sharing Fund

The above-named fund invests directly in high-impact projects to support farmers in developing countries to conserve crop diversity in their fields and assist farmers and breeders globally to adapt crops to our changing needs and demands. Ensuring sustainability of the efforts is the major focus, hence, the Fund focuses on capacity-building of developing countries, unfortunately not the least developed countries, which could be a great reservoir of varied CWRs, to enhance the exchange of information, and to make appropriate technology available to conserve and sustainable use of the vast diversity.

The above Table 18.2 summarizes some of the key aspects in this context.

The Benefit sharing Fund supported the collection of, in excess of 360 accessions of traditional varieties and CWRs, and the characterization for useful traits of more than 2200 accessions of varieties held on farm and gene banks. Over 360 traditional and locally adapted accessions of 11 different crops threatened by genetic erosion, including CWRs, were collected and stored in gene banks as a safety backup. Of these, the best performing in terms of climate adaptation, pest and disease resilience, and yields were actively disseminated and promoted among farmers, thus enhancing their continued use through on-farm management (Table 18.2).

Pre breeding—its objective and potential role: A means to pre-empt the recurrence of a massive crop destruction by a single pathogen/insect pest, as in the case of rice blast, brown plant hopper or wheat rust, which has brought about untold disaster, for example, as what happened in the Punjab State in India, the "cradle of green revolution", on the elite dwarf rice and wheat varieties, introduced into India, is the diversification of the genetic base of crops. Such diversifications can be achieved through pre breeding, the collaborative endeavor undertaken by the germplasm curator and plant breeder. This has been described by the Global Partnership Initiative for Plant Breeding Capacity Building (GIPB) as "all the activities designed to identify desirable characteristics and/or genes from unadapted materials which cannot be used directly in breeding populations, and to transfer these traits to an intermediate set of materials which plant breeders can further use in producing new crop varieties for the benefit of farmers". What is needed is a combined effort between the plant breeder and gene bank manager to chalk out a strategy to utilize combined resources and expertise to achieve a common goal with the following objectives:

1. Broadening the genetic base, to minimize vulnerability
2. Identify beneficial traits in wild materials and move those genes which control these traits into breeding populations to be more readily accessed by the plant breeders

Why the need for a public-private partnership in pre breeding?
Pre breeding is a slow, and also, a very costly exercise, requiring long-term investment of resources. The expanding role of the private sector in plant breeding in many countries has been paralleled with decreasing resources in the public sector, which historically undertook most pre commercial breeding and released the results to the entire farming and plant breeding community, without exclusivity. India is a classic example, where the Indian Council of Agricultural Research, ICAR, headquartered in New Delhi, undertook such a task, which benefited the nation's entire farming community enormously. In fact, during the hey day of the green revolution starting mid-sixties lasting up to early eighties, when it had a steep fall with plateauing crop yields, due to the breakdown of many elite crop varieties, primarily of rice and wheat, due to insect pest and disease attack, soil degradation, groundwater pollution etc., due to unbridled chemical fertilizer use, none in India had heard of the private sector getting into the seed business. All that has now changed, for the worse, and to the economic detriment of the poor and marginal farmer, who is at the mercy of private seed companies, which sell the seed at an enormous price. The classic case is of the genetically modified cotton seed peddled by Monsanto and its Indian arm Mahyco (Maharashtra Hybrid Seed Company). Seed business is now a very lucrative business, and, the best example is that of the genetically modified seeds. The following box summarizes the contours of pre-breeding:

Pre breeding posits itself as a crucial and unique step between PGRFA conserved in collections, landraces, underutilized species, and CWR and their use by breeders. The gap is dealing with the lack of characterization and evaluation data that remains a serious constraint to the use of many collections. Significant attention is needed to pre breeding to enhance the use of PGRFA. As pre breeding is being carried out, the resulting materials are expected to have merit to be included in conventional breeding programs aimed at producing new crop varieties for farmers. It is best viewed as the interface between gene banks and breeding which includes activities which would normally be not part of the task of either plant breeders or of gene bank managers. The vision is integrative which captures the absolute need for total involvement, skills, and tools of gene bank managers, plant breeders and, the farming community, at large.

Numerous public-private partnerships have in recent years undertaken focused pre-breeding programs. However, it has proved difficult to attract adequate resources into pre breeding issues, because of its long-term nature and the shorter-term character of most funding sources. At the United Nations Conference on Sustainable Development, Rio de Janeiro, Brazil, on June 21, 2012, the Rio Six-Point Action Plan for the ITPGRFA (http://www.planettreaty.org/content/rio-six-point-action-plan-2012) was adopted by consensus at the Second High-level Roundtable on the ITPGRFA. As one of the six points, it calls for stakeholders "to promote a public-private partnership for pre breeding". The objective is to promote a partnership, directly managed by the partners, to ensure food security and economic development, by delivering top-quality pre bred materials to breeders in the public and private sectors.[1]

The emerging reality necessitates fostering of closer mutually beneficial relationships between this sector and the public counterpart that traditionally bred and disseminated superior crop varieties. This need becomes even more compelling in view of the germplasm accessions which constitute the basis for genetic improvement which are held by the public sector institutions. Strengthening the efforts of, and partnerships between public and private sector breeders to conserve and use PGRFA is essential. Additionally, participatory breeding and selection, as well as participatory research in general, with farmers and farming communities need to be strengthened and recognized more broadly as appropriate ways of achieving the sustainable and long-term conservation and use of PGRFA. The partnership will provide a forum for consideration of all issues important for pre breeding, including the following:

1. Establishing the agreed upon objectives and priorities
2. Support for the mobilization of resources for agreed upon activities and programs
3. The role of Intellectual Property management in public-private partnerships, with identification of the best practices and model clauses. This is the most vexatious aspect of the entire plan. This happens to be so, because, there is an all pervasive fear, also substantiated by practical evidence, that there is a subtle, yet true, and undeniable, underlying objective, hidden from public glare, that the west quite often poaches on the rare and unique plant wealth of the third world. The

---

[1] Report of the Technical Consultation to Promote Public-Private Partnerships for Pre-Breeding. http://www.planttreaty.org/sites/default/files/gb5i17_Prebreeding.pdf.

siphoning of rare rice germplasm from India, during the heyday of the green revolution to IRRI, facilitated by unscrupulous Indian agricultural scientists, which led to the rewarding of the latter, through dubious awards and monetary gain through appointments to high positions in international institutions dealing with agricultural research, which brought the manipulators much pecuniary benefit, is an example in the history of Indian agriculture. The details of this was published under the title "The Great Gene Robbery" by a leading Indian journalist Mr Claude Alvares of Goa, on March 23, 1986, in the most popular and widely read weekly magazine of India, "The Illustrated Weekly", which shook the scientific world, globally, and in India, in particular, and, led to the disastrous course of Indian agriculture, culminating in the fall of the green revolution due to contamination of the native blood of rice varieties with the alien blood (Taiwan) in the much-publicized "dwarf, elite" varieties of rice, which succumbed to severe insect pest (Brown Plant Hopper) and disease (Rice Blast), devastating thousands of acres of rice, in the State of Punjab, India, the "cradle of Indian green revolution". The rare rice varieties of States like Kerala, in southern India, practically disappeared (Rice to the Challenge, The Hindu January 13, 2018), due to this poaching and clandestine export overseas.

4.   Facilitate the integration of technologies in pre-breeding

The partnership itself should not seek to undertake pre-breeding, but, provide a policy and support forum to strengthen pre breeding, with the aim of promoting the widest possible access to high-potential, stabilized, pre commercial materials.

Reference

Mercer KL, Perales HR (2010) Evolutionary response of landraces to climate change in centres of diversity. Evol Appl 3:480–493

Chapter 19
Economics of CWR Under Climate Change

Despite difficulties involved in precisely quantifying the undeniable economic value of CWR, it can never be underestimated. CWR is a treasure house of genetic material which can provide immense benefits to agriculture when integrated into cultivated crop varieties. CWR species have been used in crop breeding to introduce genes into cultivated varieties to impart traits, such as, drought and frost tolerance, resilience to insect pests and diseases, and, specific abiotic stress like soil acidity and salinity. The following Table 19.1 incorporates examples of CWR gene introduction into cultivated crop species to enhance yield/and or combat specific biotic or abiotic stresses.

Several of these investigations, cited below in Table 19.1, rely on simple calculations and approximations to estimate the value of genetic contributions from CWRs. Several other estimates of the economic value of CWR genetic contributions exist. In addition to the studies discussed above, one of the most comprehensive attempts to analyze the economic value of CWR in breeding is that undertaken by Prescott-Allen and Prescott-Allen's (1986): "The First Resource: Wild Species in the North American Economy". The authors estimate that the annual value of a genetic resource (that is, that provided by a gene from a CWR) will only be worth some proportion of the value of that crop, at the most. Investigating the successful use of CWR in plant breeding, Prescott-Allen and Prescott-Allen (1986) estimate the value of the CWR contribution as being f% of p% of the value of the crop, p% being the proportion of the crop supplied by cultivars with wild germplasm, and f% being the proportion of the value of those cultivars accounted for by factors controlled by wild germplasm. The following Table 19.2 presents their estimates of the annual value of genetic contributions provided by CWR to a number of crops produced within and imported into the USA.

A recent and sophisticated analysis of CWR economics is Hein and Gatzweiler's (2006) investigation of the value of wild coffee (*Coffea arabica*) genetic resources. The authors assess both the potential costs and benefits of the use of wild coffee genetic material in breeding programs for enhanced coffee cultivars (possessing increased pest and disease tolerance, low caffeine contents, and increased yields), instead of just estimating the proportion of the value of a crop that is attributable to

© Springer Nature Switzerland AG 2019
K. P. Nair, *Combating Global Warming*, Springer Climate,
https://doi.org/10.1007/978-3-030-23037-1_19

Table 19.1 Examples of quantifiable benefits of CWR gene incorporation. The data pertains to past estimates of the economic value of CWR in the published dollar figure and amount adjusted to US Dollar parity (2012)

Parameter investigated	Figure, US $ (million/billion)	Adjusted figure, US $ (million)	Reference
Annual benefits from disease resistance introgressed from wild wheat species	50 million	107 million	Witt (1985)
Annual contributions of CWR to US economy from domestic and imported sources	340 million	712 million	Prescott-Allen and Prescott-Allen (1986)
Annual contribution of genes from wild tomato species *Lycopersicon chmielewski*	8 million	16 million	Iltis (1988)
Annual contributions of CWR to US economy	20 billion	28.61 billion	Pimentel (1997)
Annual contribution of CWR to world economy	115 billion	164.5 billion	Pimentel (1997)
Net present value of wild coffee genetic resources	1.458 billion	1.660 billion	Hein and Gatzweiler (2006)
Annual value of increase in crop productivity because of CWR genetic contributions	1 billion	1.686 billion	NRC (1991)
Annual contributions of wild sunflower gene pool	267–384 million	273–392 million	Hunter and Heywood (2011)
Annual contributions of a wild tomato species providing a 2.4% increase in solids content	250 million	255 million	Hunter and Heywood (2011)
Current value of CWR genetic contributions	68 billion	N.A.	PwC (2013)
Potential value of CWR genetic contributions	196 billion	N.A.	PwC (2013)

Note: N.A. = Not Available

Table 19.2 Estimated annual value of genetic contributions provided by CWR to USA agriculture in 1986 (1986 US$ parity and in 2012 US$ parity)

Crop	Percentage of total (%)	Total annual value of contributions, 1986 US $ parity	Total value of contributions, 2012 US $ parity
Sugarcane	34.8	119, 400, 000	250, 120, 000
Sunflower	25.8	88, 500, 000	185, 390,000
Cacao	14.3	49, 000, 000	102, 650, 000
Bread wheat	10.3	35, 300, 000	73, 950, 000
Oil palm	9.4	32, 300,000	67, 660, 000
Hops	3.5	11, 900, 000	24, 930, 000
Sugar beet	0.9	3.0–3.25 million	6.28–6.81 million
Tobacco	0.5	1, 900, 000	3, 980 000
Oats	0.4	0.6–2.3 million	1.26–4.82 million
Bermuda grass	0.1	400, 000	837, 900
Total[a]	342.3–344.25 million	717.1–721.1 million	

[a]Prescott-Allen and Prescott-Allen (1986) also included potato, tomato, bell pepper, strawberry, cotton, tulip, sweet clover, alfalfa, Spanish iris, smooth brome, high bush blueberry, and lettuce as crops with breeding contributions from wild species, but, were not able to estimate the value of these contributions

CWR genetic material. Using a 30-year discounting period to compare the potential present and future costs and benefits of the breeding program, the authors estimate the net present value of wild coffee genetic resources in Ethiopia at US $ 1458 and US $ 420 million (using discount rates of 5% and 10% respectively). This investigation is a step forward in that it accounts for both the costs and benefits of using CWR in breeding, instead of just estimating the annual economic benefits in terms of gross production value. This method helps to differentiate between CWR species which are easy and those which are difficult to use in breeding by factoring in the costs of use. This investigation also provides a methodology to estimate the value of at-risk genetic resources, such as, wild coffee forests in highland Ethiopian forests.

The most recent valuation of CWR genetic resources for Kew's Millennium Seedbank was conducted by Pricewaterhouse Coopers (PwC) LLP (PwC 2013). Their analysis measured the current and potential economic value of the CWR of potato, rice, wheat, and cassava in terms of the additional gross production value provided by the genetic contributions from CWR species, an approach similar to that which is used by Prescott-Allen and Prescott-Allen (1986). Current value was calculated as being the economic benefit over the useful economic life of traits in current crops, and potential value was approximated as the economic benefits from future improved crops into perpetuity. PwC (2013) then extrapolated the average increase in value because of CWR from these crops to the rest of the 29 priority crop gene pools chosen by Kew[3], as well as maize, sugarcane, and soybean. The results are summarized in the following Table 19.3.

Table 19.3 PwC estimates of the current and potential value of CWR genetic contributions (US $ 2013 parity)

Details	Current value (US $) (billion)	Potential value (US $) (billion)
Wild gene pools of potato, rice, wheat, and cassava	25	73
Wild gene pools of 29 priority crops[2]	42	120
Wild gene pools of 29 priority crops, maize, sugarcane, and soybean	68	196

It is difficult to precisely estimate the value of CWRs, and, many of the investigations discussed above rely on simple, back-of-the envelope calculations which are likely to be inaccurate. These CWR valuation studies are limited by the quality of the data they rely on, as well as the models they use, and most of the estimates are dated, though they do provide a valuable foundation for future attempts to estimate the economic value of CWRs. However, in the context of climate change, it is necessary to investigate the economic value of CWRs in a more comprehensive manner.

19.1 Quantifying the Economic Value of CWR Under the Impact of Climate Change

It is essential that a broader framework is adopted to quantify the economic benefits which flow from CWR to fully understand the CWR value, as of the present, and in the future, under the impact of climate change. Most of the scientists who approached this issue, as discussed earlier in this chapter, attempted to estimate the genetic contributions from CWR by analyzing the approximate increase in the gross production value of the crop (either through increased yields or improved quality) because of the introgressed wild genetic material. The following equation, depicted by the following Fig. 4 presents an example of expanded framework to measure the economic value of CWR under the impact of climate change.

Under this expanded framework, one can observe that the farmer benefits from any genetic trait provided by CWR through breeding, which either increases 1a (the gross production value of the crop) through a yield increase or quality improvement, or decreases 2a, 3a, or 4a (the costs of inputs such as water, pesticides and herbicides, and fertilizer, respectively). The "farmer" value equation can be expressed as follows:

$$1a - (2a + 3a + 4a) = \text{Value to farmer} \tag{19.1}$$

The above equation can be pictorially depicted as follows:

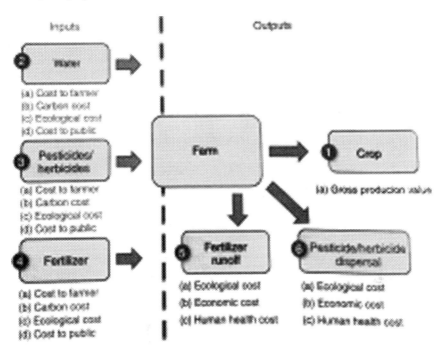

Fig. 19.1 A conceptual framework to measure the economic value of CWR

The above equation is already an improvement on the models which look at the value of the CWR genetic contribution solely as a portion of the gross production value of the crop. Indeed, a wild gene for drought tolerance/resistance or tolerance/resistance to an insect pest/plant disease might not increase the gross production value of the crop produced, at all, but, it can reduce the need for water, pesticides and/or herbicides, and thus can decrease the input costs for the farm. Similarly, a wild gene that serves the function of decreasing a crop's need for nutrients, such as nitrogen or phosphorus, might reduce the amount of fertilizer needed to achieve the same yield, again leading to lower costs and an economic benefit for the farmer would accrue. These genetic improvements might also lead to consumer surplus if the farmer sells his/her crop for less on the market because of his/her reduced operating costs. The potential economic benefit in terms of cost reductions (2a, 3a, and 4a) could be estimated by determining how much less water, fertilizer, or pesticides the improved crop variety needs to produce the same yield in relation to the next best cultivar, and then multiplying this yield advantage by the cost of the input and the area of farm land in which the improved variety is cultivated. However, there are also a range of other external costs ("externalities") which impact the public (if not the farmer). For instance, many agricultural inputs have substantial carbon costs (2b, 3b, and 4b). Fertilizers, pesticides, and herbicides are all produced, transported, and applied using fossil fuels. The production of fertilizers leads to between 35.73

and 857.54 kg of carbon dioxide emitted per 1000 kg produced, depending on the fertilizer. And fertilizer application resulted in carbon emissions of an estimated 12.35 kg of carbon dioxide emitted per hectare in 1990. And the use of nitrogenous fertilizers like urea leads to a substantial emission of nitrous oxide (N_2O) into the atmosphere which has a five-fold higher heat-trapping ability than carbon di oxide, thus leading to increased global warming. And, this is another great environmental cost of the green revolution, where unbridled use of chemical fertilizers, especially nitrogenous like urea, to boost grain yield of crops like rice and wheat has been the order of the day. The pesticide production similarly leads to, according to a 2002 investigation, between 4702.38 and 5177.52 kg carbon dioxide emitted per 1000 kg produced, and pesticide application led to an estimated 2.54 kg of carbon dioxide emitted per hectare in 1990. Irrigation also results in carbon emissions, as water is often pumped from groundwater reserves and transported, sometimes, long distances to reach farmers' fields. On an average, 266.48 kg of carbon dioxide are emitted for every hectare of farmland irrigated in the USA (West and Marland 2002). Thus, if breeding with CWR species can help reduce the need for these inputs, then, the value of their genetic contributions should be calculated as including potentially significant carbon savings—savings that have a dollar value in carbon markets. This finding has vast environmental importance in the introgression of CWRs in conventional plant breeding.

The above-discussed framework provides an improved method to determine the economic value of CWR. However, it is important to include climate change in the equation, described above, in particular, when determining what the potential value of a crop variety containing CWR genes might be in the future. Global warming is predicted to decrease water availability in parts of the world, which will lead to the price escalation of water. When input costs in farming are escalating by the day, especially in the case of both fertilizer and water, it is but important to determine the positive contribution in maintaining and/or enhancing crop productivity through the introgression of a CWR gene into the usually cultivated cultivar. Hence, an improved cultivar containing CWR genes which reduce water needs of the crop is most likely to produce more economic benefits under future changed climate than it will to day. Similarly, an improved crop variety containing CWR genes which make it more drought tolerant is likely to be more valuable in future changed climate, due to global warming, in which droughts are projected to be more frequent than they are at present. It is also important to realize that a CWR might provide resilience to a particular crop in times of emergencies like famine and/or extreme climatic aberrations. If CWR genetic material can help avert famine by boosting or maintaining agricultural production during droughts or epidemics of insect pests and plant diseases, both predicted to be more recurrent in the climate change scenario of the future, CWR species can be credited with providing an additional value associated with the prevention of malnutrition, death, and potential conflicts which would have occurred due to food scarcity. The classical example is of *Oryza nivara* which staved off a famine, in many countries when the rice crop was getting devastated by the grassy stunt virus disease during the seventies (Brar and Khush 1997). It is difficult to estimate what this value is, or make predictions on how costly a famine would have been in the absence of the improved crop. However, it is clear that any genetic contributions of CWR species which increase the resilience of agriculture to the stresses of climate change and which help mitigate and/or pre-empt famine have value, in addition to that estimated using the expanded valuation framework described above. The value of CWR can be measured in many more ways than just as a proportion of the gross production value of a crop. It extends to the inputs a farmer buys and a number of other externalities related to these inputs as well. Additionally, the CWR generates a great public good, which provides greater justification for the use of these wild species in agricultural research.

References

Brar DS, Khush GS (1997) Alien introgression in rice. Plant Mol Biol 35:35–47

Hein L, Gatzweiler H (2006) The economic value of coffee (Coffea arabica) genetic resources. Ecol Econ 60:176–185

Hunter D, Heywood V (2011) Crop wild relatives: a manual of in situ conservation, 1st edn. Earthscan, London, UK

Iltis HH (1988) Serendipity in the exploration of biodiversity. What good are weedy tomatoes? In: Wilson EO (ed) Biodiversity. National Academy Press, Washington, DC, pp 98–105

NRC (National Research Council) (1991) Managing global genetic resources: the US national plant germplasm system. National Academy Press, Washington, DC

Prescott-Allen C, Prescott-Allen R (1986) The first resource: wild species in the North American economy. Yale University Press, New Haven, CT, p 529

PricewaterhouseCoopers LLP (PwC) (2013) Crop wild relatives: a valuable resource for crop development. http://www.pwc.co.uk/sustainability-climate-change/publications/understandingthe-value-of-seeds.jhtml. Last accessed 6 Nov 2014

West TO, Marland G (2002) A synthesis of carbon sequestration, carbon emissions, and net carbon flux in agriculture: comparing tillage practices in the United States. Agric Ecosyst Environ 91:217–232

Witt SC (1985) Biotechnology and genetic diversity. California Lands Project, San Francisco, CA

Chapter 20
Conservation Economics of CWR

While analyzing the economic value of CWR, the costs of conserving them should also be taken into consideration, because, without foolproof conservation, the inherent value of a CWR might simply be lost. CWR can be conserved in situ, in the wild, or ex situ in gene banks. One advantage in maintaining CWR in situ is that CWR conserved thus continue to evolve and can generate novel and potentially more adaptive traits as the climate changes. For example, wild wheat and barley populations in Israel have been shown to have shifted their flowering dates significantly earlier in the season from 1980 to 2008 to escape the adverse effect of increased drought brought about by global warming (Nevo 2011). CWR populations in the wild also tend to contain greater genetic diversity than samples of seeds contained in a gene bank. Hunter and Heywood (2011) have written a manual detailing information on in situ conservation.

The in situ conservation costs of CWR can vary as much as do the habitats in which these plants are found, and depend on the opportunity cost of using the land in other ways, such as, for firewood production, farmland, building houses, and other developmental activities. It may be relatively cheap to flag "CWR hotspots" for conservation in the desert, for example, wild wheat and barley species found in the Middle East, and relatively expensive to set aside, for example, forestland containing populations of wild fruit tree relatives, as in Indonesia, where rainforest is prized for timber and conversion to oil palm plantations. The annual costs of effective field-based terrestrial conservation may vary from less than US $ 0.1/km^2 to US $ 1 million/km^2, depending on the location and habitat (Balmford et al. 2003). In situ conservation may also coexist with other uses of the land which do not negatively affect CWR populations, potentially helping to reduce conservation costs.

When ex situ conservation is done with CWR in gene banks, it provides additional security and increases the availability of these genetic resources for use in breeding. Many CWR accessions (or samples) can be centralized in a single location, assessed for valuable traits and bred with cultivated varieties to produce improved crop varieties. CWR accessions can also be regenerated and shared between institutions, often facilitated by their inclusion in Annex 1 of International Treaty on Plant

© Springer Nature Switzerland AG 2019
K. P. Nair, *Combating Global Warming*, Springer Climate,
https://doi.org/10.1007/978-3-030-23037-1_20

Table 20.1 Ex situ
conservation costs for
selected CWR gene pools

Details (wild gene pool)	Gene bank	Annual recurring conservation cost per accession in US $ (2010)
Barley (*Hordeum vulgare*)	ICARDA	5.65
Bean (*Phaseolus vulgaris*)	CIAT	19.48
Rice (Oryza sativa)	IRRI	21.27
Sorghum (*Sorghum bicoolr*)	ICRISAT	25.50
Groundnut (*Arachis hypogaea*)	ICRISAT	38.22
Cassava (*Manihot utilissima*)	CIAT	71.88

Genetic Resources for Food and Agriculture. Examples of ex situ conservation costs for selected CWR gene pools is given in the above Table 20.1.

The monetary value of conserved CWR can be categorized into the following three categories (Garming et al. 2010):

1. CWRs which are being used in breeding in the present have a current use value by virtue of their contributions (through breeding) to improved crop varieties
2. CWR accessions which have been found to possess valuable genetic material through evaluation activities have an expected "future use value" in addition to their current use value, as estimated for wild coffee genetic resources in Ethiopia by Hein and Gatzweiler (2006)
3. While it is often unclear what value CWR samples may have for breeding, these resources may have high option value, in particular, if they are at risk of extinction outside of where they are conserved. This option value refers to the possibility of finding useful genes in conserved CWR species in the future to cope with unforeseeable shocks and challenges

If the option value of the CWR is ignored, it is likely that wild genetic diversity of potentially significant future value may be lost. Though CWR accessions or populations of CWR in the wild which have not been used or found to have useful traits or genes have low current value, and a low predicted future use value, they may yet contain valuable traits which are, as of now, undiscovered. The option value associated with conserving CWRs is inherently uncertain, difficult to quantify, but conserving the greatest possible diversity of CWR germplasm may be the best strategy ensuring that wild plant genetic resources are available to adapt to contain future climate-related shocks.

References

Balmford A, Gaston KJ, Blyth S, James A, Kapos V (2003) Global variation in terrestrial conservation costs, conservation benefits, and unmet conservation needs. Proc Natl Acad Sci USA 100(3):1046–1050

Garming H, Roux N, Van den Houwe I (2010) The impact of the musa international transit centre—review of its services and cost effectiveness and recommendations for rationalization of its operations. Biodiversity International, Montpellier, France

Hein L, Gatzweiler H (2006) The economic value of coffee (Coffea arabica) genetic resources. Ecol Econ 60:176–185

Hunter D, Heywood V (2011) Crop wild relatives: a manual of in situ conservation, 1st edn. Earthscan, London, UK

Nevo E (2011) Triticum. In: Kole C (Ed), Wild crop relatives: genomics and breeding resources. Cereals. (Chapter 10). Springer, Berlin, Heidelberg, pp 407–456

Chapter 21
The Millennium Seed Bank—Their Conservation Roles and the Svalbard Global Seed Vault

According to FAO statistics (FAO 2010), more than 2500 botanical gardens and 1750 gene banks exist worldwide. In addition to the conservation of species and their genetic diversity, these structures help to increase public awareness of the importance of preserving biodiversity and contribute to the raising of funds for conservation projects. Additionally, they are also important centers of research and development into plant taxonomy, genetics and seed ecology. Seed saving for crops is an ancient tradition in almost all cultures, especially in countries like India, and, as old as the time of origin of farming, about 10,000 years ago. In fact, in many farming communities in India, immediately after harvest, a portion of the harvest is kept separately for seed purposes, and preserved until the following cropping season. All that has changed with the advent of the green revolution where the hybrid seeds are accessed from seed agencies, either public or privately owned. In remote corners of India, seeds of some of the crops like rice have been saved over millennia, and, they have assumed great significance in the context of the genetically modified seeds, where agribusiness interests have poached on them. Odisha State, in India, is the granary of these ancient rice seed stocks, some of which were pirated by vested interests, during the heyday of the green revolution, in lieu of the offer of plum jobs overseas in international organizations, to these vested interests in India, as a quid pro quo exercise.

Organized collection and conservation of crop genetic resources in the form of seed samples first started with the Russian botanist and plant breeder Nicolai Vavilov and the American geneticist and plant breeder Harry Harlan in the first part of the twentieth century (Vavilov 1992; Harlan 1995). Seed banks are now found all over the world and are a repository of genetic material from both wild and domesticated plant species. In general, gene banks have an explicit mandate to conserve economically important plant genetic resources for food and agriculture (PGRFA), including the wild relatives of crop plants, which have potentially useful characteristics for crop improvement. The RBG Kew Millennium Seed Bank (MSB) of the Royal Botanic Gardens, Kew (RBG Kew) and the Svalbard Global Seed Vault (SGSV) are two major global facilities in modern ex situ conservation. A pictorial representation of both is given (Figs. 21.1 and 21.2).

© Springer Nature Switzerland AG 2019
K. P. Nair, *Combating Global Warming*, Springer Climate,
https://doi.org/10.1007/978-3-030-23037-1_21

Fig. 21.1 Millennium seed
bank building, Wakehurst
place, Sussex, UK

Fig. 21.2 Frozen entrance
of the Svalbard global seed
vault, Norway

These two conservation projects bring together the economic and public awareness raising aspects of plant conservation in an extraordinary and successful manner. While the MSB is housed in the Wellcome Trust Millennium Building (WTMB), located at RBG Kew's garden of Wakehurst Place in West Sussex, UK, the SGSV is located at the arctic archipelago Svalbard, at latitude 78° N, between mainland Norway and the North Pole. The SGSV is located inside a mountain just outside Norwegian settlement Longyearbyen. Under the National Heritage Act 1983, RBG Kew is a nondepartmental public body with exempt charitable status. From the MSB RBG Kew's manages the International Millennium Seed Bank Partnership (1990) (MSBP). The MSB is the physical seed collection and the activities carried out around these, while the MSBP is the network of partners across the world working with MSB. The SGSV is managed in partnership by the Norwegian Ministry of Agriculture and Food (NMAF), the Nordic Genetic Resource Center (NordGen) and the Global Crop Diversity Trust (2004) (the Trust) (http://www.croptrust.org/). NordGen is responsible for the management of seed deposits, and the storage and the costs are jointly funded by the Trust and the Norwegian government. NordGen is a public regional institute supported by the governments of the five Nordic countries and the Trust, an independent international organization based in Bonn, Germany. The management of the SGSV is overseen by an International Advisory Council consisting of international technical and policy experts with representatives from the depositing gene banks, the United Nations FAO, CGIAR, and the Governing Body of the International Treaty on Plant Genetic Resources for Food and Agriculture (ITPGRFA). The legally responsible authority for SGSV is the NMAF.

21.1 Project Activities of MSBP and SGSV on CWR

There is a common interest between both the organizations, for the global improvement of crops. CWRs, as discussed earlier, are an important, but, underutilized resource to improve crop varieties, through yield increase, and tolerance, and/or resistance to both biotic and abiotic stresses. The global crop conservation strategies commissioned by the Trust indicate that the CWRs of the majority of the crops investigated are underrepresented in ex situ collections. Several crop expert surveys have been done to estimate the gaps in the global ex situ collections of cultivars of the most important crops. While it is estimated that only 5% of the domestic gene pool of wheat, rice, maize, and potato remains to be banked, gaps in ex situ collections of crops' primary, secondary, and tertiary gene pools are still very large (Fowler and Hodgkin, 2004). Since SGSV is a backup site for conventional gene banks, the relatively poor representation of CWRs in global ex situ collections is also reflected in the collections at Svalbard. The following Table 21.1 presents an overview of the CWRs (as defined according to Crop Wild Relatives and Climate Change (2012) Online resource, www.cwrdiversity.org) of selected crops conserved at the SGSV (Table 21.1).

Table 21.1 Accession number safety duplicated in the SGSV for five very important crops, their CWR, and for the whole genus (May 2013)

Crop name	Crop accessions	CWR accessions	Crop genus accessions
Rice (*Oryza sativa*)	137, 061	7, 224[a]	145, 698
Barley (*Hordeum vulgare*)	41,403	2, 826	61, 390
Sorghum (*Sorghum bicolor* ssp.*bicolor*)	40, 533	125	40, 695
Maize (*Zea mays* ssp.*mays*)	25, 657	29	32, 822
Common bean (*Phaseolus vulgaris* ssp. *vulgare*)	30, 901	1, 302	35, 230

Note Taxa included in the CWR category is determined according to Crop Wild Relatives and Climate Change (2012). Online resource www.cwrdiversity.org
[a]Includes 3767 *Oryza glaberrima*

The various gene pools and CWRs of barley (*Hordeum vulgare* subsp. *vulgare*) are well represented in the SGSV. The genus *Hordeum* has a natural distribution in most temperate areas in the world. The progenitor of barley, *Hordeum* vulgare, subsp.*spontaneum,* which belongs to the primary gene pool is, in particular, well represented from the area of origin in the Middle East.

One of the important projects among those aimed at adapting agriculture to combat climate change is titled "Adapting Agriculture to Climate Change" project by Dempewolf and his co-researchers (Dempewolf et al. 2014), funded by the Trust and RBG, Kew. It's objective is to collect, protect, and select a portfolio of plants, with the characteristics required to adapt the world's most important food crops to climate change. The collection program systematically targets CWR in each of these crops, with a specific focus on species related to twenty nine crops listed below, which are of utmost importance to global food security. They are:

African rice, Alfalfa, Apple, Bambara groundnut, Banana, Barley, Carrot, Chickpea, Common bean, Cowpea, Egg plant, Faba bean, Finger millet, Grasspea, Lentil, Lima bean, Oat, Pea, Pearl millet, Pigeon pea, Plantain, Potato, Rice, Rye, Sorghum, Sunflower, Sweet potato, Vetch, and Wheat. The material collected through this project will be duplicated in the first instance at the MSB, but, later through the CGIAR regular backup system to the SGSV. The seeds from the MSB will also be sent directly to pre breeders for each crop to directly stimulate evaluation and characterization of the material. The seed collections housed at the MSB and SGSV are of enormous value to conservation. When one considers the social and economic importance/potential that seeds, such as those of the CWRs, have for humanity, it is unsurprising that they attract so much of keen public interest.

21.2   The Millennium Seed Bank (MSB)

A new era in seed conservation was initiated by the launching of the MSB at the beginning of the current millennium. The layout of the MSB building is designed to engage and educate the public of the activities of the seed bank. The Orange Room at the center is a public space surrounded by laboratories, where visitors can witness all technical aspects of seed conservation. Both artists and art lovers are attracted to the MSB, because, they are inspired by the beauty, complexity, and ingenuity of the seeds, themselves. The breath-taking architectural design of seeds has been showcased in the book Seeds: Time Capsules of Life (Kesseler et al. 2006), and some of these exquisite electron micrographs are on display in the Orange Room. Hanging from the glass ceiling is a 3 m long fiberglass structure of a devil's claw seed pod, created by Tony Gibas, and dotted throughout the gardens are giant seed sculptures by Tom Hare. These and other exhibitions raise the public profile of seed conservation and show how closely art and nature are interlinked.

The MSB website (www.kew.org/msbp/) is a public facing platform which features articles, image galleries, and blogs which describe the activities of the seed bank, and update readers on the current aims, projects, and partnerships of the MSB.

While the primary intention of MSB is to analyze these data for predictive patterns which support seed conservation operations, it is likely that a wide variety of users outside the project will find the data valuable for many purposes. It is important to note that school children from all over the United Kingdom regularly visit the seed bank and the education department provides learning programs and tours for the students.

References

Dempewolf HD, Eastwood RJ, Guarino L, Khoury CK, Muller JV (2014) Adapting agriculture to climate change: a global 10 year project to collect and conserve crop wild relatives. Agroecol Sustain Food Sys 28:369–377

FAO (2010) The second report on the state of the world's plant genetic resources for food and agriculture. Italy, FAO, Rome, p 370

Fowler C, Hodgkin T (2004) Plant genetic resources for food and agriculture: assessing global availability. Ann Rev Env Resour 29:143–179

Harlan JR (1995) The living fields. Our agricultural heritage. Cambridge University Press, Cambridge, UK. New York, USA, Melbourne, Australia

Kesseler R, Stuppy W, Papadakis A (2006) Seeds: time capsules of life. Firefly Books

Vavilov NI (1992) Origin and geography of cultivated plants. Cambridge University Press, Cambridge, UK

Chapter 22
The Svalbard Global Seed Vault

The SGSV was officially inaugurated in February 2008. It offers a unique possibility for exposure concerning issues of conservation of plant genetic resources. Situated in a remote and exotic location in the arctic, it has attracted great attention and triggered the imagination of persons involved in future development. In earlier times, after inauguration, it was looked upon as a futuristic and future-oriented installation with a thrilling content. People often wonder whether this is "A Doom's Day Vault" for a remote future, promoting speculation and theories of conspiracy, which has inspired many writers and filmmakers, to make their pieces, sometimes in a bizarre fashion! There is no diminished interest in SGSV, in fact, the attention is directed to the actual seed content, conditions in the SGSV and its management. It is a symbol of domesticated biodiversity conservation, and, in this respect serves as a background for information and raising awareness and knowledge for a number of important issues and putting them into a general context, concerning the following issues:

1. Global biodiversity status
2. Conservation of genetic diversity of plant genetic resources for food and agriculture (PGRFA)
3. Genetic resources utilization in pre breeding and conventional plant breeding
4. Agriculture and horticulture global situation in both developing and industrialized countries
5. Concentrating on a number of urgent issues of importance for the global future, which includes climate change and its consequence on food security, overpopulation and the urgent need for more food, for which the central elements are genetic resources, including the CWRs, and plant breeding. Wherever required, supportive measures targeting food security are also addressed, especially with regard to the economic use of external inputs in crop production, and sustaining it at optimum level, such as, chemical fertilizers, water, and pesticides. It is in this context that a revolutionary soil management technique developed by the author of this book, globally known as "The Nutrient Buffer Power Concept For Sustainable Agriculture" (Nair 1996, 2010, 2013) assumes significance.

© Springer Nature Switzerland AG 2019
K. P. Nair, *Combating Global Warming*, Springer Climate,
https://doi.org/10.1007/978-3-030-23037-1_22

During 2009–2012, 500 requests were received by SGSV concerning:

1. Information on conservation and utilization of plant genetic resources, and, in particular, the management of the SGSV and its seed content
2. Request for lectures on SGSV
3. Requests for visiting SGSV

The SGSV stipulates a very strict policy for visits, on account of safety and security reasons, set up by the Trust, the Norwegian Government, and NordGen, which ensures that only a few visits are permitted, manly for politicians, (ministers of agriculture, rural affairs, environment, foreign affairs, education and science) policymakers and media. As the SGSV does not have a permanent staff on the site, only a limited number of visits can be permitted in a year. Among the visitors are many from the UN, advisors to politicians, and representatives of influential organizations. Though interest shown by the general public, tourists, scientists, students, and commercial enterprises is great, these categories are disallowed visits. However, there are films, for example, "The Back—Up Copy" made by Snow Ball Film, books, articles in magazines, and newspapers, TV and radio programs, etc., through which information reaches a large audience around the world. One specific category of audience which has shown immense interest in the SGSV is the various art projects, photo exhibitions, art books, different installations, and, specifically, projects on architecture are attracted by the beauty of the SGSV, the art installations, light and the inspiring content. To promote further interest in the SGSV, and creating more awareness in biodiversity conservation, following activities may be taken up:

1. Lectures, attendance in various meetings, events, and exhibitions
2. Seminars, and conferences with MSB and SGSV in focus
3. Development of more material for exhibitions
4. Making more direct invitations to politicians and policymakers around the world
5. Enthusing primarily the children, the makers of the future world.

References

Nair KPP (1996) The buffering power of plant nutrients and effects on availability. Adv Agron 57:237–287
Nair KPP (2010) "The Nutrient Buffer Power Concept"—a revolutionary soil management technique. In: Proceedings of the 19th world congress of soil science, Brisbane, Australia
Nair KPP (2013) The buffer power concept and its relevance in African and Asian soils. Adv Agron 121:416–516

Printed in the United States
By Bookmasters